THE FERTILITY OF THE UNFIT

By

William Allan Chapple

dition et mise en page : Culturea (Hérault, 34)
etrouvez notre catalogue sur http://culturea.fr/
ontact : infos@culturea.fr
nprimé en Allemagne par Books on Demand Gmbh
esign typographique et layout : Derek Murphy et Reedsy
BN :9791041985463
ate de parution : février 2024

INTRODUCTION.

Biology is the Science of Life. It seeks to explain the phenomena of all life, whether animal or vegetable. Its methods are observation and experiment. It observes the tiny cell on the surface of an egg yolk, and watches it divide and multiply until it becomes a great mass of cells, which group off or differentiate, and rearrange and alter their shapes. It observes how little organs unfold themselves, or evolve out of these little cell groups—how gradual, but how unvarying the change; how one group becomes a bone, another a brain, another a muscle, to constitute in three short weeks the body of a matured chick. Those little tendons like silken threads, that run down those slender pink legs to each and every toe, and move its little joints so swiftly that we hardly see them—that little brain, no bigger than a tiny seed, in which is planted a mysterious force that impels it to set all those brand-new muscles in motion, and to dart after a fly with the swiftness of an arrow—all this wondrous mechanism, all this beauteous structure, all this perfection of function, all this adaptation to environment, have evolved from a few microscopic cells in three short weeks.

Biology is the science that observes all this, and enunciates the law that the life history of this animal cell, *i.e.*, its history from a simple unicellular state in the egg, to its complex multicellular state in the matured chick, represents the history of the race to which the chick belongs. If we could trace that chicken back through all its ancestry, we would discover at different periods in the history of life upon the globe (about 100 million years, according to Haeckel) exactly the stages of development we found in the life history of the chick, and arrive at last at a primordial cell.

What is true of the chick is true of all life. This is the law of evolution. It is true of all plant and animal life; it is true of man as an individual; it is true of his mind as well as of his body; it is true of society as an aggregation of individuals. As men have evolved from a lower to a higher, a simple to a complex state, so they are still evolving and rising "on stepping stones of their dead selves to higher things."

Natural selection, or the survival of the fittest, is one of the processes by which evolution takes place. According to this law, only the fittest survive in the struggle for life. Darwin was led to this discovery on reading Malthus's thesis regarding the disproportion between the rates of increase in population and food, and the consequent struggle for existence.

All living organisms require food and space. The power of multiplication in plants and animals is so great that food or space is sooner or later entrenched upon, and then commences this inevitable struggle for existence. In this struggle for life, the individuals best able to conform to their environment, *i.e.*, the best able to resist adverse circumstances, to sustain hardships, to overcome difficulties, to defend themselves, to outstrip their fellows, in short, to harmonise function with environment, survive. These propagate their kind according to the law of heredity. Variations exist in the progeny, and the individuals whose variations best adapt them to their environment are the fittest to, and do, survive.

In a state of nature the weaklings perish. If man interferes with this state of nature in the lower animals, he may make a selection and cultivate some particular attribute. This is artificial selection, and is best exemplified in the experiments with pigeons. Pasteur saved the silk industry of France, and perhaps of the whole world, by the application of this law of artificial selection. The disease of silkworms, known as Pebrine, was spreading with ruinous rapidity in France. Pasteur demonstrated that the germ of the disease could be detected in the blood of affected moths by the aid of the microscope. He proved that the eggs of diseased moths produced unhealthy worms, and he advised that the eggs of each moth be kept apart, until the moth was examined for germs. If these were found, the eggs were to be burned. Thus the eggs of unhealthy moths were never hatched, and artificial selection of healthy stock stamped out a disease, and saved a great industry.

Each individual plant in the struggle for life has only itself to maintain. In the higher forms of animal life, each animal has its offspring as well as itself to maintain. In a state of nature, that is in a state unaffected by man's rational interference, defective offspring and weaker brethren were the victims of the inexorable law of natural selection. When Christ gave *his* reply to the question, "Am I my brother's keeper?" the defective and the weakling became the special care of their stronger brother. They constituted thenceforth The Fit Man's Burden. The work a man has to do during life, in order to support himself, is the unit of measurement of the burden he has to bear. Many factors in modern times have helped to reduce that work to a minimum. The invention of machinery has multiplied his eyes, his hands, his feet; and one man can now produce, for his own maintenance and comfort, what it took perhaps a score of men to produce even a century ago. Man's disabilities from incidental and epidemic disease have been immeasurably reduced by modern sanitation, and the teaching and practice of preventive medicine. Agricultural chemistry has made the soil more productive, and manufacturing arts have aided distribution as well as production.

All the departments of human knowledge have been placed under contribution to man's necessity, and longer life, better health, and more food and clothing for less work, are the blessings on his head to-day.

While the burden has been lessened by the industrial and scientific progress of the last half century, it has been augmented by the fertility of the unfit; and the maintenance in idleness and comfort of the great and increasing army of defectives constitutes the fit man's burden. The unfit in the State include all those mental and moral and physical defectives who are unable or unwilling to support themselves according to the recognised laws of human society. They include the criminal, the pauper, the idiot and imbecile, the lunatic, the drunkard, the deformed, and the diseased. We are now face to face with the startling fact that this army of defectives is increasing in numbers and relative fertility.

Consider what a burden is the criminal. Every community is more or less terrorised by him; our property is liable to be plundered, our houses invaded, our women ravished, our children murdered. To restrain him we must build gaols, and keep immense staffs of highly paid officials to tend him in confinement, and watch him when he is at liberty. Notwithstanding these, crime is rife, and is rapidly increasing. Says Douglas Morrison:—"It is perfectly well known to every serious student of criminal questions, both at home and abroad, that the proportion of habitual criminals in the criminal population is steadily on the increase, and was never so high as it is now.... The population under detention in reformatory institutions is increasing more rapidly than the growth of the community as a whole, and, as far as it is possible to see, the juvenile population in prisons is doing the same thing." Havelock Ellis, Boies, and McKim, all corroborate this testimony. "Among the three or four millions of inhabitants of London, one in every five dies in gaol, prison, or workhouse."

All these defectives are prolific, and transmit their fatal taints. "In a certain family of sixteen persons, eight were born deaf and dumb, and one at least of this family transmitted the defect as far as the third generation." ("Heredity and Human Progress.") A murderer was the son of a drunkard; of three brothers, one was normal, one a drunkard, and the third was a criminal epileptic. Of his three paternal uncles, one was a murderer, one a half idiot, and one a violent character. Of his four cousins, sons of the latter, two were half idiots, one a complete idiot, and the other a lunatic.

There is an agricultural community of about 4000 in the rich and fertile district in the valley of Artena, in Italy, who have been thieves, brigands, and assassins since 1155 A.D. They were outlawed by Pope Paul IV., in 1557, but they still live and flourish in their crime, the victims of a criminal inheritance. The ratio of homicides in Italy and Artena is as 9 to 61; of assault and battery as 34 to 205; of highway robbery as 3 to 145; of theft as 47 to 111. Professor Pellman, of Bonn University, has traced the careers of a large number of defectives, and shown their cost to the State. Take this example:—A woman who was a thief, a drunkard, and a tramp for forty years of her life, had 834 descendants, 709 of whom were traced; 106 were born out of wedlock, 142 were beggars, and 64 more lived on charity. Of the women, 181 lived disreputable lives. There were in the family 76 convicts, 7 of whom were convicted of murder. In 75 years, this family cost their country in almshouses, trials, courts, prisons, and correctional establishments about £250,000. The injury inflicted by this one family on person and property was simply incalculable.

In New Zealand, the ratio of those dependent upon the State, or on public or private support, has gone up from 16.86 per thousand of population, over 15 years of age in 1878, to 23.01 in 1901. The ratio of defectives, including deaf and dumb, blind, lunatics, epileptics, paralytics, crippled and deformed, debilitated and infirm, has gone up from 5.4 per thousand, over fifteen years, in 1874, to 11.4 in 1896, declining slightly to 10.29 in 1901. The ratio of lunatics has gone up from 1.9, in 1874, to 3.4 in 1901. This is the period of the most rapid and persistent decline in the New Zealand birth-rate; and, coincident with this period, the marriage-rate went down from 8.8 per thousand in 1874, to 5.8 in 1886, and then gradually rose to 7.83 in 1901. The number of weekly rations (Parkes's standard), purchasable by the average weekly wages of an artisan in Wellington province, has gone up from 11 to 16.5 between the years 1877 and 1897. In other words, the price of food and the rate of wages in 1897 would enable an artisan to fill 5½ more mouths than he could have done at the rates prevailing in 1877.

Notwithstanding the development of civilising, Christianising, and educational institutions, crime, insanity, and pauperism are increasing with startling rapidity. The true cause is to be found deep down in biological truth. Society is breeding from defective stock. The best fit to produce the best offspring are ceasing to produce their kind, while the fertility of the worst remains undisturbed. The most striking demographical phenomenon of recent years is the declining birth-rate of civilised nations. In Germany the birth-rate has fallen from 40 to 35 per thousand of the population; in England from 35 to 30; in Ireland from 26 to 22; in France from 26 to 21; and in the United States from 36 to 30 during the last twenty years; while, in New Zealand, it has declined from 40.8, in 1880, to 25.6, in 1900. In Australia there were 47,000 less births in 1899 than would have occurred under the rates prevailing ten years ago.

There is a consensus of opinion among demographists that this decline is due to the voluntary curtailment of the family in married life. Prudence is the motive, and self-restraint the means by which this curtailment is made possible. But prudence and self-restraint are the characteristic attributes of the best citizens. They are conspicuous by their absence in the worst; and it is a matter of common observation that the hopelessly poor, the drunken and improvident, the criminal and the defective have the largest families, while those in the higher walks of life rejoice in smaller numbers. The very qualities, therefore, that make the social unit a law-abiding and useful citizen, who could and should raise the best progeny for the State, also enable him to limit his family, or escape the responsibility of family life altogether; while, on the other hand, the very qualities which make a man a social burden, a criminal, a pauper, or a drunkard—improvidence and defective inhibition—ensure that his fertility will be unrestrained, except by the checks of biological law. And it now comes about that the good citizen, who curtails his family, has the defective offspring of the bad citizen thrown upon his hands to support; and the humanitarian zeal, born of Christian sentiment, which is at flood-tide to-day, ensures that all the defectives born to the world shall not only be nursed and tended, but shall have the same opportunities of the highest possible fertility enjoyed by their defective progenitors.

A higher and nobler human happiness is attainable only through social evolution, and this comes from greater freedom of thought, from bolder enquiry, from broader experience, and from a scientific study of the laws of causation. What "is" becomes "right" from custom, but with our yearnings for a higher ideal, sentiment slowly yields to the logic of comparison, and, often wiping from our eyes the sorrows over vanishing idols, we behold broader vistas of human powers, possibilities, duties, and destiny.

As the proper study of mankind is man, influenced wholly by a desire to be useful to a society to which I am indebted for the pleasures of civilised life, I offer this brief volume as a comment on a phase of the social condition of the times, and as my conclusions regarding its interest for the future.

CHAPTER I.

THE PROBLEM STATED.

A century has passed since Malthus made his immortal contribution to the supreme problem of all ages and all people, but the whole aspect of the population question has changed since his day. The change, however, was anticipated by the great economist, and predicted in the words:—"The history of modern civilisation is largely the history of the gradual victory of the third check over the two others". The third check is moral restraint and the two others vice and misery.

The statistics of all civilized nations show a gradual and progressive decline in the birth-rate much more marked of recent years. In Germany, between the years 1875 and 1899, it has diminished from 40 to 35.9 per thousand of the population. In England and Wales, it dropped from 35 to 29.3 during the same time; in Ireland, from 26 to 22.9; in France, from 26 to 21.9; in the United States of America (between the years 1880 and 1890) the decline has been from 36 to 30; while in New Zealand it gradually and persistently declined from 40.8 in 1880 to 25.6 in 1900.

During the period, 1875-1890, the rapid strides made in industry and production have been unparallelled in the history of the world, Wealth has accumulated on all sides, and production and distribution have far outrun the needs and demands of population. To-day food is far more abundant, cheaper, and therefore more accessible to all classes of the people than it was 50 years ago, and coincident with this rapid and abundant increase in those things which go to supply the necessities, the comforts, and even the luxuries of life, there has been a constant and uniform decline in the birth-rate, and this decrease is even more conspicuous in those nations in which the rate of production has been most pronounced. It would even be true to say that the birth-rate during recent years is in inverse proportion to the rate of production.

At first sight this might appear to falsify the law of population enunciated by Malthus. Malthus maintained that population tended to increase beyond the means of subsistence; that three checks constantly operated to limit population—vice, misery, and moral restraint: vice, due largely to diseased conditions, misery, due to poverty and want, and moral restraint due to a dread of these. I shall show later that nothing has been said or written to add to or take away from the truth and force of these great principles, but, that the moral restraint of Malthus has been practised to an extent, and in a direction of which the great economist never dreamt. By moral restraint in the limitation of families Malthus meant only delayed marriage. In so far as men and women abstained from, or delayed their marriage, on the ground of inability to support a family, they fulfilled the law, and followed the advice of Malthus. Continence without the marriage bond was assumed; incontinence was classed with another check vice.

Contrary to the expectations arising out of the famous progressions, wealth and production have increased and the birth-rate has decreased. It is the purpose of this work to show what are the causes that have led to this decline, that those causes are not equally operative through all classes of the people, and that the chief cause of the decline of the birth-rate is the desire on the part of both sexes to limit the number they have to support and educate. The considerations that lead up to, and, to some extent, justify this desire, will be discussed later.

The fact remains that an increasingly large number of people have come to the conclusion that the burden and responsibility of family obligations limit their enjoyments in life, their ambition, and even their scope for usefulness, and have discovered, through the spread of physiological information, means by which marriage may be entered upon without necessarily incurring these responsibilities and limitations.

It is the knowledge of these physiological laws and the practice of rules arising out of that knowledge, that account for the declining birth-rate of civilized nations.

If it be true that the birth-rate is controlled by a voluntary effort on the part of married people to limit their families, and that that effort implies self restraint and self denial, it would not be too much to claim that those most capable of exercising self-control and with the strongest motives for such exercise, are those most responsible for the declining birth-rate, and that those with least self-control and the fewest motives for exercising the control they have, are most likely to have the normal number of children.

It has already been suggested, that the desire to limit families is due to a consciousness of responsibility on the part of prospective parents. They realise the stress of competition in the struggle for existence, they are anxious for their own pecuniary and social stability, and even more anxious that the children, for whose birth they are responsible, should be provided with the necessities and comforts of life which health and development require. They are eager, too, that their children should be equipped with a good education, and thus be given a fair advantage in the race of life.

To the great mass of people this is possible only when the numbers of the family are limited. As the numbers of the family increase, the difficulties of clothing and feeding and educating increase, and each member is the poorer for every birth, and in this sense an increasing birth-rate is a cause of poverty. The sense in which poverty causes a high birth-rate will be dealt with later on.

It will be readily conceded, that those actuated by the motives just considered, those with the keenest sense of responsibility in life, those capable of exercising the self-restraint which family limitation requires, constitute the best type of citizens in any community. From such the State has good reason to expect the best stock.

It is one purpose of this work to show that this class, which can and should produce the best in the largest numbers, is being overwhelmed with the burden of supporting an ever-increasing number of incapables, and, largely in consequence of this increasing burden and responsibility, are unwilling to produce, because they are unable adequately to support their own kind.

There is a class in every large community, whose sense of responsibility in life is at zero, whose self-control is substituted by the law and its sanctions, and whose modes and habits of life are little better than those of the lower animals. Their appetites are stronger, their desires, though fewer, are more intense, and their self-control less easily and less frequently exerted than those in the highest planes of life.

In the first place then they have less desire to limit their families, and less power to exercise the self-restraint that is necessary to do so. Less sense of responsibility is attached to the rearing of a family, whilst the education of their children gives them little or no concern. They entertain no ambition that members of their family should compete in the struggle for social status. Their instincts and their impulses are their guide in all things. They marry early, and procreation is unrestrained except by the hardships of life.

This constitutes a numerous class in every large community, and includes the criminal, the drunkard, and the pauper, and many defectives such as epileptics and imbeciles. Now all these propagate their kind. The checks to the increase of this class, are the checks which are common to the lower animals, and which were elaborated in his first essay by Malthus. They are vice and misery.

If it were not for moral restraint (not the limited restraint of Malthus, delayed marriages simply), but restraint in the wider sense, within as well as without the marriage bond, and including all artificial checks to conception, these two checks, vice and misery, would absolutely control the population of the world.

The mind of man has added to the checks which control increase in the lower animals, a new check, which applies to, and can be exercised only by himself, and the problem is, how far will misery and vice as checks to the population be eliminated, and moral restraint take their places? And if this restraint must control and determine the population of the future how far will its exercise affect the moral and mental evolution of the race?

If moral restraint with the consequent limitations of families is the peculiar characteristic of the best people in the state, and the absence of this characteristic expressing itself in normal fertility is peculiar to the worst people of the state, the future of the race may be divined, by reference to the history of the great nations of antiquity.

An accumulating amount of evidence shows that society is face to face with this grave aspect of the population question. The birth-rate of the unfit is steadily maintained. Improved conditions of life increase the number that arrive at maturity and enter the procreative period, so that not only are defectives born into the world at a constant rate, but sanitary laws and a growing impatience with the sufferings of the poor, tend so to improve their conditions of life, as to increase their birth-rate and their chances of arriving at adult life.

Shortly stated then, the problem that society has to solve is this,—The birth-rate is rapidly declining amongst the most fit to produce the best offspring, while it is steadily maintained amongst the least fit, so that the relative proportion of the unfit born into the world is annually increasing.

What should be the State's attitude to this problem, and how it should attempt to solve it will be discussed in detail in a subsequent chapter. Let it suffice to say now, that the right of the State to interfere directly with the limitation of families amongst the best classes would find few advocates amongst reformers.

The right of the State to say, however, that the criminal, the drunkard, the diseased, and the pauper, shall not propagate their kind should be stoutly maintained by all rational men.

Most of the nations of history have recognized the gravity of the population question, but they were mostly concerned with the tendency of the numbers in the State to increase beyond the means of subsistence, instead of the tendency to degeneration as it now concerns us.

CHAPTER II.

THE POPULATION QUESTION.

Births, deaths, and migration are the factors which make up the population question.

The problem has burned in the minds of all great students of human life and its conditions.

Aristotle says (Politics ii. 7-5) "The legislator who fixes the amount of property should also fix the number of children, for if they are too many for the property, the law must be broken." And he proceeds to advise (ib. vii. 16-15) "As to the exposure and rearing of children, let there be a law that no deformed child shall live, but where there are too many (for in our State population has a limit) when couples have children in excess and the state of feeling is adverse to the exposure of offspring, let abortion be procured."

The difficulty of over-population was conspicuous in the minds of Aristotle and Plato, and these philosophers both held that the State had a right and a duty to control it.

But some States were almost annihilated because they were not sufficiently populous, and Aristotle attributes the defeat of Sparta on one celebrated occasion to this fact. He says:—"The legislators wanting to have as many Spartans as they could, encouraged the citizens to have large families, and there is a law at Sparta, that the father of three sons should be exempt from military service, and he who has four, from all the burdens of the State. Yet it is obvious that if there were many children, the land being distributed as it is, many of these must necessarily fall into poverty."

The problem in the mind of the Greek philosophers was this. Over-population is a cause of poverty; under-population is a cause of weakness. Defectives are an additional burden to the State. How shall population be so regulated as to established an equilibrium between the stability of the State, and the highest well-being of the citizens?

The combined philosophy of the Greeks counselled the encouragement of the best citizens to increase their kind, and the practice of the exposure of infants and abortion.

A century of debate has raged round the name of Malthus, the great modern analyst of the population problem. He published his first essay on population in 1798, a modest pamphlet, which fed so voraciously on the criticism supplied to it, that it developed into a mighty contribution to a great social problem, second only in time and in honour to the work of his great predecessor in economic studies, Adam Smith.

Malthus's first essay defined and described the laws of multiplication as they apply only to the lower animals and savage man. It was only in his revised work, published five years later, that he described moral restraint as a third check to population.

Adverse criticism had been bitter and severe, and Malthus saw that his first work had been premature. He went to the continent to study the problem from personal observation in different countries. He profited by his observation, and by the writings of his critics, and published his matured work in 1803.

The distinguishing feature about this edition was the addition of moral restraint as a check, to the two already described, vice and misery.

Malthus maintained that population has the power of doubling itself every 25 years. Not that it *does* so, or *had done* so, or *will do so,* but that it is *capable* of doing so, and he instanced the American Colonies to prove this statement.

One would scarcely think it was necessary to enforce this distinction, between what population has done, or is doing, and what it is capable of doing. But when social writers, like Francesco Nitti, urge as an argument against Malthus's position that, if his principles were true, a population of 176,000,000 in the year 1800 would have required a population of only one in the time of our Saviour, it is necessary to insist upon the difference between *increase* and the *power of increase*.

One specific instance of this doubling process is sufficient to prove the *power of increase* possessed by a community, and the instance of the American Colonies, cited by Malthus, has never been denied.

A doubling of population in 25 years was thus looked upon by Malthus as the normal increase, under the most favourable conditions; but the checks to increase, vice, misery, and moral restraint are operative in varying degrees of intensity in civilized communities, and these may limit the doubling to once in 50, or once in 100 years, stop it altogether, or even sweep a nation from the face of the earth.

The natural increase among the lower animals is limited by misery only, in savage man by vice and misery only, and in civilized man by misery, vice, and moral restraint.

Misery is caused by poverty, or the need of food or clothing, and is thus proportionate to the means of subsistence. As the means of subsistence are abundant, misery will be less, the death-rate lower, and *caeteris paribus* the birth-rate higher. The increase will be directly proportional to the means of subsistence.

Vice as a check to increase, is common to civilized and savage man, and limits population by artificial checks to conception, abortion, infanticide, disease, and war. The third check, moral restraint, is peculiar to civilized man, and in the writings of Malthus, consists in restraint from marriage or simply delayed marriage.

Bonar says, "Moral restraint in the pages of Malthus, simply means continence which is abstinence from marriage followed by no irregularities."

These checks have their origin in a need for, and scarcity of food,—food comprising all those conditions necessary to healthy life. The need of food is vital and permanent. The desire for food, immediate and prospective, is the first motive of all animal activity, but the amount of food available in the world is limited, and the possible increase of food is estimated by Malthus at an arithmetical ratio.

Whether or not this is an accurate estimate of the ratio of food increase is immaterial. Malthus's famous progressions, the geometrical ratio of increase in the case of animals, and the arithmetical ratio of increase in the case of food, contain the vital and irrefutable truth of the immense disproportion between the power of reproduction in man and the power of production in food.

Under the normal conditions of life, the population tends constantly to press upon, and is restrained by the limits of food. The true significance of the word *tends* must not be overlooked, or a similar fallacy to that of Nitti's will occur, when he overlooked the significance of the term "power to multiply." It is perfectly true to say, that population *tends* to press upon the limits of subsistence, and unrestrained by moral means or man's reason actually does so.

Some social writers appear to think that, if they can show that production has far outstripped population, that, in other words, population for the last fifty years at least has *not* pressed upon the limits of food, Malthus by that fact is refuted.

Nitti says, "But now that statistics have made such great progress, and the comparison between the population and the means of subsistence in a fixed period of time is no longer based upon hypothesis, but upon concrete and certain data in a science of observation it is no longer possible to give the name of law to a theory like that of Malthus, which is a complete disagreement with facts. As our century has been free from the wars, pestilences and famines which have afflicted other ages, population has increased as it never did before, and, nevertheless, the production of the means of subsistence has far exceeded the increase of men."

And later he says "Malthus's law explains nothing just as it comprehends nothing. Bound by rigid formulas which are belied by history and demography, it is incapable of explaining not only the mystery of poverty, but the alternate reverses of human civilization."

Nitti's conclusions are based largely on the fact that while food supplies have become abundant and cheap, birth-rates have steadily and persistently declined.

No-one who has studied the economic and vital statistics of the last half century can fail to be impressed with the change that has come over the relative ratios of increase in population and food.

Bonar says, "The industrial progress of the country (France) has been very great. Fifty years ago, the production of wheat was only half of what it is to-day, of meat less than half. In almost every crop, and every kind of food, France is richer now than then, in the proportion of 2 to 1. In all the conveniences of life (if food be the necessaries) the increased supply is as 4 to 1, while foreign trade has become as 6 to 1."

In a remarkable table prepared by Mr. F.W. Galton, and quoted by Mr. Sydney Webb in "Industrial Democracy," it is clearly shown, that, while the birth-rate and food-rate (defined as the amount of wheat in Imperial quarters, purchased with a full week's wages) gradually increased along parallel lines between 1846 and 1877, the former suddenly decreased from 36.5 per thousand in 1877 to 30 per thousand in 1895, the latter increasing from .6 to 1.7 for the same period.

The remarkable thing about the facts that this table so clearly discloses is that with a gradual increase of the means of subsistence from 1846 to 1877 there is also a gradual increase in the proportion of births to population. But at the year 1877 there, is a very sudden and striking increase in food products, and the purchasing power of the people coincides exactly with a very sudden and striking decrease in the birth-rate of the people. The greater the decrease in the birth-rate, the greater the increase in the people's purchasing power. Now, what has brought about this change in the ratios of increase in population and in food respectively?

Some serious factor, inoperative during the thirty years prior to 1877 must have suddenly been introduced into the social system, to work such a marvellous revolution during the last twenty years.

Some economic writers find it easy here to discover a law, and declare that the birth-rate is in inverse ratio to the abundance of food.

Other economic writers of recent date attribute this great change in ratio of increase to economic causes. Only a few find the explanation in biological laws.

Herbert Spencer is the champion of the biological explanation of a decreasing birth-rate.

With the intellectual progress of the race there is a decadence of sexual instinct. In proportion as an individual concentrates his energies and attention on his own mental development, does the instinct to, and power of, generation decrease.

It may be true, it certainly is true, that if an individual's energies are concentrated in the direction of development of one system of the body, the other systems to some extent suffer. A great and constant devotion to the development of the muscular system will produce very powerful muscles, and great muscular energy, with a strong tendency to, and pleasure in exercise. It is true also, that time and energy are monopolized in this creation of muscle, and that less time and energy are available for mental pursuits and mental exercise.

Up to a certain point muscular exercise aids mental development, but beyond that point concentration of effort in the direction of muscular development starves mental growth.

On the other hand, if the education and exercise of the mind receive all attention, the muscular system will suffer, and to some extent remain undeveloped. Or generally, one system of the body can be highly developed only at the expense of some other system, not immediately concerned.

It is true that the more an individual concentrates his efforts on his own intellectual development, the more his sexual system suffers, and the less vigorous his sexual instincts.

And the converse of this is also true, for examples of those with great sexual powers are numerous.

In plant life, this same law is also in operation. If one system in a plant, the woody fibre for instance, takes on abundant growth, the fruit is starved and is less in quality and quantity, and *vice versa*.

But to what extent does this affect fertility? Sexual power and fertility are not synonymous terms.

The vast profusion of seed in plant and animal life, would allow of an enormous reduction in the amount produced, without the least affecting fertility. Even admitting the application of Spencer's law to sexual vitality, and allowing him to claim that, with the progress of "individuation," there is a decline in sexual instinct, would the fertility of the race be affected thereby?

To have any effect at all on the birth-rate, the instinct would have either to be killed or to be so reduced in intensity as to stop marriage, or to delay it till very late in life.

When once marriage was contracted sexual union once in every two years, would, under strictly normal conditions, result in a very large family.

For according to Mr. Spencer's theory, it is the instinct that is weakened not the power of the spermatozoa to fertilize.

Evidence is wanting, however, to show that there is a decrease in the sexual power of any nation.

France might be flattered to be told that her low birth-rate is due to the high intellectual attainments of her people, and that the rapidly decreasing birth-rate is due to a rapid increase of her intellectual power during recent years.

Ireland and New Zealand would be equally pleased could they believe that their low, and still decreasing birth-rate is due to the lessening of the sexual instinct, attendant upon, and resulting from a high and increasing intellectual power and activity.

The fact is, that the sexual instinct is so immeasurably in excess of the maximum power of procreation in the female, that an enormous reduction in sexual power would require to take place before it would have any effect on the number of children born.

The number of children born is controlled by the capacity of the human female to bear children, and one birth in every two years during the child-bearing period of life is about the maximum capacity.

A moderate diminution in the force of the sexual instinct might lead to a decrease in the marriage rate, but it would require a very serious diminution bordering on total extinction of the instinct to exert any serious effect on the fecundity of marriage.

All that can be claimed for this theory of population is, that, reasoning from known physiological analogies, we might expect a weakening of the desire for marriage, coincident with the general development of intellect in the race.

There are as yet no facts to prove that such weakening has taken or is taking place, nor are there facts to prove that population has in any way suffered from this cause.

If such a law obtained, and resulted in a diminished birth-rate, the future of the race would be the gloomiest possible. An inexorable law would determine that there could be no mental evolution, for the best of the race would cease to propagate their kind. All who would arrive at this standard of mental growth would become barren. And against this there could be no remedy.

One of the main contentions of this work is that the best have to a large extent ceased to propagate their kind, but it is not maintained that this is the result of a biological law, over which there is no control. It can be safely claimed that to Malthus's three checks to population—vice, misery, and moral restraint, the demographic phenomena of a century have added no other. The third check, however, moral restraint, must be held to include all restraint voluntarily placed by men and women on the free and natural exercise of their powers of procreation.

Malthus used the term "moral" in this connection, not so much in relation to the *motive* for the restraint, but in relation to the result, viz., the limitation of the family. The "moral restraint" of Malthus meant to him, restraint from marriage only, chiefly because of the inability to support a family. It implied marriage delayed until there was reasonable hope that the normal family, four in number, could be comfortably supported, continence in the mean time being assumed. Bonar interpreting Malthus says that impure celibacy falls under the head of "vice," and not of "moral restraint."

To Malthus, vice and misery, as checks to population, were an evil greatly to be deplored in civilized man, and not only did he declare that moral restraint obtained as a check, but he also declared it a virtue to be advocated and encouraged in the interest of society, as well as of the individual.

His moral restraint was delayed marriage with continence. He trusted to the moral force of the sexual passion in a continent man to stimulate to work, to thrift, to marriage; to work and save so that he may enter the marriage state with a reasonable prospect of being able to support a wife and family.

Malthus never anticipated the changes and developments of recent years. He advised moral restraint as a preventive measure in the hope that vice and misery, as checks would be superseded, and that no more would be born into the world than there was ample food to supply. He believed that moral restraint was the check of civilized man, and as civilization proceeded, this check would replace the others, and prevent absolutely the population pressing upon the limits of subsistence.

He saw in moral restraint only self-denial, constant continence, and entertained not a doubt, that the generative instinct would be cheated of its natural fruit. The passion for marriage is so strong (thought Malthus) that there is no fear for the race; it cannot be over-controlled.

The gratification of the sexual instinct, and procreation were the same thing in the mind of Malthus.

But this is not so.

A physiological law makes it possible, in a large proportion of strictly normal women, for union to take place without fertilisation. If it were possible to maintain an intermittent restraint in strict conformity with this law, it would control considerably the population of the world.

It is easier to practice intermittent than to practice constant restraint.

It is just here that Malthus failed to anticipate the future. Malthus believed that "moral restraint" would lessen the marriage rate, but would have no direct effect on the fecundity of marriage.

A man would not put upon himself the self-denial and restraint, which abstinence from marriage implied, for a longer period than he could help.

The greater the national prosperity, therefore, the higher the birth-rate. But prosperity keeps well in advance of the birth-rate; in other words, population, though it still *tends* to, does not actually *press* upon the food supply.

If the moral restraint of Malthus be extended so as to include intermittent moral restraint within the marriage bond, then, under one or other, or all of his three checks, vice, misery, and moral restraint, will be found the explanation of the remarkable demographic phenomena of recent years.

Misery will cover deaths from starvation and poverty, the limitation of births from abortion due to hardship, from deaths due to improper food, clothing, and housing; and emigration to avoid hardship.

Vice will cover criminal abortions, limitation of births from venereal disease, deaths from intemperance, etc., and artificial checks to conception. Malthus included artificial checks of this kind under vice, though they have some claim to be considered under moral restraint. But the question will be referred to in a later chapter.

Moral restraint will cover those checks to conception, voluntarily practised in order to escape the burden and responsibility of rearing children—continence, delayed marriage, and intermittent restraint.

No other checks are directly operative.

Misgovernment and the unequal distribution of wealth and land affect population indirectly only, and can only act through one or other or all of the checks already mentioned.

CHAPTER III.

DECLINING BIRTH-RATE.

It is not the purpose of this work to follow any further the population problem so far as it relates to deaths and emigration. Attention will be concentrated on births, and the influences which control their rates.

A rapid and continuous decline in the birth-rate of Northern and Western Europe, in contravention of all known biological and economic laws, has filled demographists with amazement.

A table attached here shows the decline very clearly. According to Parkes, the usual food of the soldier may be expressed as follows: —

Articles.	Daily quantity in oz. av.
Meat	12.0
Bread	24.0
Potatoes	16.0
Other vegetables	8.0
Milk	3.25
Sugar	1.33
Salt	0.25
Coffee	0.33
Tea	0.16

Total	65.32
Butter	2.4 — (Moleschott.)

The New Zealand Official Year Book gives the following as the average prices of food for the years mentioned:—

		1877		1887		1897		1901	
		s.	d.	s.	d.	s.	d.	s.	d.
Bread	per lb.	0	2¼	0	1¾	0	1½	0	1½
Beef	per lb.	0	5¼	0	3½	0	3	0	5
Mutton	per lb.	0	4	0	2¾	0	2	0	4½
Sugar	per lb.	0	5¾	0	3	0	2½	0	2¾
Tea	per lb.	3	0	2	3	2	0	1	10
Butter (fresh)	per lb.	1	3	1	0	0	8	0	11
Cheese (col'n'l)	per lb.	0	10	0	5¾	0	6	0	6
Milk	per qt.	0	4½	0	3	0	3	0	3½

The official returns give the average daily wage for artisans for the years 1877, 1887, 1897, and 1901 as 11s., 10s. 6d., 9s. 9d., and 10s. 3d., respectively.

The weekly rations (the standard food supply for soldiers—Parkes's) purchaseable by the weekly wages for these years respectively are 11.1, 14.3, 16, and 12.4; *i.e.*, the average weekly wage of an artisan in constant employment in 1877 would purchase rations for 11.1 persons, in 1887 for 14.3 persons, in 1897 for 16 persons, and in 1901 for 12.4 persons.

Up to the year 1877, the birth-rate in England and Wales conformed to the law of Malthus, and kept pace with increasing prosperity; but, after that year, and right up to the present time, the nation's prosperity has gone on advancing at a phenomenal rate *pari passu* with an equally phenomenal decline in the number of births per 1000 of the population.

Now, it is a remarkable coincidence that in this very year, 1877, the Neo-Malthusians began to make their influence felt, and spread amongst all classes of the people a knowledge of preventive checks to conception.

People were encouraged to believe that large families were an evil. A great many, no doubt, had already come to this conclusion; for there is no more common belief amongst the working classes, at least, than that large families are a cause of poverty and hardship. And this is even more true than it was in the days of the Neo-Malthusians, for then child and women labour was a source of gain to the family, and a poor man's earnings were often considerably augmented thereby.

The uniform decrease of the birth-rate is a matter of statistics, and admits of no dispute. It has been least rapid in the German Empire, and most rapid in New Zealand.

With the declining birth-rate the marriage-rate must be considered.

Malthus would have expected a declining birth-rate to be the natural result of a declining marriage-rate, and a declining marriage-rate to be due to the practice of moral restraint, rendered imperative because of hard times, and a difficulty in obtaining work, wages, and food.

Given the purchasing power of a people, Malthus would have estimated, according to his laws, the marriage-rate, and, given the marriage-rate, he would have estimated the birth-rate.

But anticipations in this direction, based on Malthus's laws, have not been realised. The purchasing power of the people we know has enormously increased; the marriage-rate has not increased, it has, in fact, slightly decreased; but the birth-rate per marriage, or the fecundity of marriage, has decreased in a remarkable degree.

In "Industrial Democracy," by Sydney and Beatrice Webb, the following occurs:—"The Hearts of Oak Friendly Society is the largest centralised Benefit Society in this country, having now over two hundred thousand adult male members. No one is admitted who is not of good character, and in receipt of wages of twenty-four shillings a week or upwards. The membership consists, therefore, of the artisan and skilled operative class, with some intermixture of the small shopkeeper, to the exclusion of the mere labourer. Among its provisions, is the "Lying-in Benefit," a payment of thirty shillings for each confinement of a member's wife.

From 1866 to 1880 the proportion of lying-in claims to membership slowly rose from 21.76 to 24.78 per 100. From 1880 to the present time it has continuously declined, until now it is only between 14 and 15 per 100.

The following table (from the annual reports of the Committee of Management of the Hearts of Oak Friendly Society, and those of the Registrar-General) shows, for each year from 1866 to 1895 inclusive, the number of members in the Hearts of Oak Friendly Society at the beginning of the year, the number of those who received Lying-in Benefit during the year, the percentage of these to the membership at the beginning of the year, and the birth-rate per thousand of the whole population of England and Wales.

HEARTS OF OAK FRIENDLY SOCIETY.

Year.	Number of Members at the beginning of each year.	Number of Cases of lying-in Benefit paid during year.	Percentage of cases paid to total Membership at beginning of year.	England and Wales: births per 1000 of the total population.
1866	10,571	2,300	21.76	35.2
1867	12,051	2,853	23.68	35.4
1868	13,568	3,075	22.66	35.8
1869	15,903	3,509	22.07	34.8
1870	18,369	4,173	22.72	35.2
1871	21,484	4,685	21.81	35.0
1872	26,510	6,156	23.22	35.6
1873	32,837	7,386	22.49	35.4
1874	40,740	9,603	23.57	36.0

1875	51,144	13,103	23.66	35.4
1876	64,421	15,473	24.02	36.3
1877	76,369	18,423	24.11	36.0
1878	84,471	20,409	24.16	35.5
1879	90,603	22,057	24.34	34.7
1880	91,986	22,740	24.72	34.2
1881	93,615	21,950	23.45	33.9
1882	96,006	21,860	22.77	33.8
1883	98,873	21,577	21.82	33.5
1884	104,339	21,375	20.51	33.6
1885	105,622	21,277	20.14	32.9
1886	109,074	21,856	20.04	32.8
1887	111,937	20,590	18.39	31.9
1888	115,803	20,244	17.48	31.2
1889	123,223	20,503	16.64	31.1
1890	131,057	20,402	15.57	30.2
1891	141,269	22,500	15.93	31.4
1892	153,595	23,471	15.28	30.5
1893	169,344	25,430	15.02	30.8
1894	184,629	27,000	14.08	29.6
1895	201,075	29,263	14.55	30.4
1896	206,673	30,313	14.67	...

In this remarkable table the percentage of births to total membership gradually rose from 21.76, in 1866, to 24.72, in 1880, and then gradually declined to 14.67 in 1896.

This is a striking instance of the fact that the decrease in the total birth-rate is due more to a decrease in the fecundity of marriage, than to a decrease of the marriage-rate.

Mr. Webb adds:—"The well-known actuary, Mr. R.P. Hardy, watching the statistics year by year, and knowing intimately all the circumstances of the organisation, attributes this startling reduction in the number of births of children to these specially prosperous and specially thrifty artisans entirely to their deliberate desire to limit the size of their families."

The marriage-rate in England and Wales commenced to decline about three years before the sudden change in the birth-rate of 1877, and continued to fall till about 1880, but has maintained a fairly uniform standard since then, rising slightly in fact, the birth-rate, meanwhile, descending rapidly.

CHAPTER IV.

MEANS ADOPTED..

There is a gradually increasing consensus of opinion amongst statisticians, that the explanation of the decrease in the number of births is to be found in the desire of married persons to limit the family they have to rear and educate, and the voluntary practice of certain checks to conception in order to fulfil this desire.

It is assumed that there is no diminution in the natural fertility of either sex. There is no evidence to show that sexual desire is not as powerful and universal as it ever was in the history of the race; nor is there any evidence to show that the generative elements have lost any of their fertilizing and developmental properties and power.

Dr. J.S. Billings in the June number of the *Forum* for 1893, says that "the most important factor in the change is the deliberate and voluntary avoidance or prevention of child-bearing on the part of a steadily increasing number of married people, who not only prefer to have but few children, but who know how to obtain their wish."

He further says, "there is no good reason for thinking that there is a diminished power to produce children in either sex."

M. Arsène Dumont in "Natalite et Democratie" discusses the declining birth-rate of France, and finds the cause to be the voluntary prevention of child-bearing on the part of the people, going so far as to say that where large families occur amongst the peasantry, it is due to ignorance of the means of prevention.

The birth-rate in none of the civilized countries of the world has diminished so rapidly as in New Zealand. It was 40.8 in 1880; it was 25.6 in 1900, a loss of 15.2 births per 1000 of the population in 20 years.

There is no known economic cause for this decline. The prosperity of the Colony has been most marked during these years.

Observation and statistics force upon us the conclusion that voluntary effort upon the part of married couples to prevent conception is the one great cause of the low and declining birth-rate.

The means adopted are artificial checks and intermittent sexual restraint, within the marriage bond, the latter tending to replace the former amongst normal women, as physiological knowledge spreads.

Delayed marriage still has its influence on the birth-rate, but with the spread of the same knowledge, that influence is a distinguishing quantity.

Delayed marriage under Malthusian principles would exert a potent influence in limiting the births, because early marriages were, and, under normal circumstances would still be, fruitful.

In the 28th annual report relating to the registration and return of Births, Marriages and Deaths in Michigan for the year 1894, it is stated that "The mean number of children borne by females married at from 15 to 19 years of age inclusive, is 6.76. For the next five year period of ages, it is 5.32, or a loss of 1.44 children per marriage, this attending an advance of five years in age at marriage."

Voluntary effort frequently expresses itself in the practice of abortion. Many monthly nurses degenerate into abortionists and practise their calling largely, while many women have learned successfully to operate on themselves.

The extent to which this method of limiting births is practised, and the absence of public sentiment against it, in fact the wide-spread sympathy extended to it, may be surmised from the facts that at a recent trial of a Doctor in Christchurch, New Zealand, for alleged criminal abortion, a large crowd gathered outside the Court, greeting the accused by a demonstration in his favour on his being discharged by the jury. A similar verdict in a similar case in Auckland, New Zealand, was greeted by applause by the spectators in a crowded Court, which brought down the indignant censure of the presiding Judge.

In New Zealand there is no oppressive misgovernment, there is no land question in the sense in which Nitti applies the term, there is no poverty to account for a declining birth-rate or to confuse the problem. There is prosperity on every hand, and want is almost unknown. And yet, fewer and fewer children, in proportion to the population, and in proportion to the number of marriages, are born into the colony every year. The only reason that can be given is that the people, though they want marriage and do marry, do not wish to bear more children than they can safely, easily, and healthfully support, with a due and ever-increasing regard for their own personal comfort and happiness. They have learned that marriage and procreation are not necessarily inseperable and they practice what they know.

CHAPTER V.

CAUSES OF DECLINING BIRTH-RATE.

The first or direct cause of this decline in the birth-rate then, is the inhibition of conception by voluntary means, on the part of those capable of bearing children.

This inhibition is the result of a desire on the part of both sexes to limit their families.

Conception is inhibited by means which do not necessitate continence, but which do necessitate some, and in many cases, a great amount of self-restraint. But how comes it, that in these days of progress and prosperity, especially in New Zealand, a desire to limit offspring should exist amongst its people, and that the desire should be so strong and so universal?

The desire for this limitation must be strong, for there is absolutely no evidence that the passion for marriage has lost any of its force; it must be extensive for the statistics show its results, and the experience of medical men bears the contention out.

While the marriage passion remains normal, offspring cannot be limited without the exercise of self-restraint on the part of both parties to the marriage compact. Artificial means of inhibiting conception, and intermittent restraint are antagonistic to the sexual instinct, and the desire for limitation must be strong and mutual to counteract this instinct within the marriage bond.

The reasons for this strong and very general desire, that marriage should not result in numerous births must have some foundation. What is it?

It cannot be poverty. New Zealand's economic experience has been one of uniform progress and prosperity. There is abundant and fertile land in these islands where droughts, floods, and famine years, are practically unknown. Blissards and destructive storms are mysterious terms. Fluctuations in production take place of course, but not such as to result in want, to any noticeable extent. There are no extremes of heat and cold, no extremes of drought and flood, no extremes of wealth and poverty. The climate is equable, the progress is uniform, the classes are at peace.

Every natural blessing that a people could desire in a country, is to be found in New Zealand. Climate, natural fertility, and production, unrivalled scenery in mountain, lake, and forest, everything to bless and prosper the present, and inspire hope in the future. Why is it that, with all this wealth, and with the country still progressing and yet undeveloped, a desire exists in the heart of the people to limit families.

The reason is social not economic, if one may contrast the terms.

Take women's attitude to the question first. Our women are well educated. A state system of compulsory education has placed within the reach of all a good education, up to what is known as the VI. or VII. Standard, and only a very few in the colony have been too poor or too rich to take advantage of it.

Most women can and do read an extensive literature, and to this they have abundant access, for even small country towns have good libraries. Alexandra, a little town of 400 inhabitants amongst the Central Otago mountains, has a public library of several thousand volumes, and the people take as much pride in this institution as in their school and church.

People move about from place to place, and it is surprising how small and even large families keep migrating from one part of the colony to another. They are always making new friends and acquaintances, and with these interchanging ideas and information.

Class distinctions have no clear and defined line of demarcation, and there is a free migration between all the classes; the highest, which is not very high, is always being recruited from those below, and from even the lowest, which is not very low.

The highest class is not completely out of sight of any class below it, and many families are distributed evenly over all the classes. A woman is the wife of a judge, a sister is the President of a Woman's Union, another sister is in a shop, and a fourth is married to a labourer.

If one of the poorer (they do not like "lower") class rises in the social scale, he or she is welcome—if one of the richer (they do not like "higher") falls, no effort is made by the class they formerly belonged to to maintain her status in order to save its dignity or repute.

In other words, there are not the hindrances to free migration between the various strata of society that obtain in other lands. Not only is that migration continually taking place, but there are very few who are not touched by a consciousness of it.

Members of the lower strata, all well educated voters, can give instances of friends, or relatives, or acquaintances, who are higher up than themselves—have "made their way," have "risen in society," have "done well," are "well off." And this consciousness inspires in all but the very lowest classes an ambition to rise.

Because it is possible to rise, because others rise, the desire to be migrating upwards soon takes possession of members of all but the lowest or poorest class, or those heavily ballasted with a large or increasing family.

The desire to rise in social status is inseparably bound up with the kindred desire to rise in the standard of comfort and ease.

Social status in New Zealand is, as yet, scarcely distinguishable from financial status. Those who are referred to as the better classes, are simply those who have got, or who have made, money. All things, therefore, are possible to everyone in this democratic colony.

There is thus permeating all classes in New Zealand a spirit of social rivalry, which shows no tendency to abate nor to be diverted. The social status of one class exerts an attractive force on the class next below.

But, apart from the influence of status, one class keeps steadily in view, and persistently strives to attain, the ease, comfort, and even luxury of the class above it.

Because the members of different grades are so migratory, there are many in one class known well to members in some class or classes below, and the ease and luxury which the former enjoy are a constant demonstration of what is possible to all.

Many who do not acquire wealth enough to make any appreciable difference in their social status, are able, through family, to improve their position. Their sons and daughters are given an University education, and by far the largest number of those entering the learned professions in New Zealand are the sons of farmers, tradespeople, and retail dealers.

The great mass of the people in our Colony are conscious of the fact that their social relations and standard of comfort, or shall one say standard of ease, are capable of improvement, and the desire to bring about that improvement is the dominant ambition of their lives.

Anything that stands in the way of this ambition must be overcome. A large family is a serious check to this ambition, so a large family must be avoided.

This desire to rise, and this dread too of incurring a responsibility that will assuredly check individual progress were counselled by Malthus, and resulted, and he said should result, in delayed marriage, lest a man, in taking to himself a wife, take also to himself a family he is unable to support.

But if this man can take to himself a wife without taking to himself a family, what then?

Men and women, in this Colony at least, have discovered that conformity to physiological law makes this possible.

A wife does not really add very much to a man's responsibility—it is the family that adds to his expense, and taxes all his resources. It is the doctor and the nurse, the food and the clothing, and the education of the uninvited ones to his home, that use up all his earnings, that keep him poor, or make him poorer.

Then there is one aspect of the question peculiar to the women themselves. Women have come to dread maternity. This is part of a general impatience with pain common to us all. Chloroform, and morphia, and cocaine, and ethyl chloride have taught us that pain is an evil.

When there was no chance of relieving it, we anæsthetised ourselves and each other with the thought that it was necessary, it was the will of Providence, the cry of our nerves for succour.

Now it is an evil, and if we must submit we do so under protest. Women now engage doctors on condition that chloroform will be administered as soon as they scream, and they scream earlier in their labour at each succeeding occasion.

Women are less than ever impressed with the sacredness and nobility of maternity, and look upon it more and more as a period of martyrdom. This attitude is in consonance with the crave for ease and luxury that is beginning to possess us.

It is, however, no new phase in human experience. It characterised all the civilisations of ancient times, at the height of their prosperity, and was really the beginning of their decay.

Women with us are more eager to limit families than are their husbands. They feel the burdens of a large family more. They are often heard to declare that, with a large family around her, and limited funds at her disposal with which to provide assistance, a woman is a slave. A large number think this, and, if there is a way out of the difficulty, they will follow that way. And they are not content to escape the hardships of life. They want comforts, and seek them earnestly. With the advent of comfort, they seek for ease, and, when this is found, they seek for luxury and social position.

Parents with us have a high ideal of what upbringing should be. Every parent wants his children to "do better" than himself. If he does not wish to make a stepping-stone of them, on which to rise to higher social things, he certainly wishes to give them such a "start in life" as will give them the best prospects of keeping pace with, or outstripping their fellows.

The toil and self-denial that many poor parents undergo, in order to give their children a good education, is almost pathetic, and is not eclipsed by the enthusiasm for education even in Scotland.

There is a shoemaker in a small digging town in New Zealand, still toiling away at his last, whose son is a distinguished graduate of our University, author of several books, and in a high position in his profession.

There is a grocer in another remote inland village whose son is a doctor in good practice. There is a baker in a little country district whose sons now hold high positions in the medical profession, one at home and the other abroad.

These facts are widely known amongst the working classes, and inspire them with a spirit of rivalry.

With regard to the general education of the people, the Registrar-General says, (New Zealand Official Year Book for 1898, page 164) "In considering the proportions of the population at different age periods, the improvement in education is even more clearly proved. It is found that, in 1896, of persons at the age-period 10-15 years, 98.73 per cent, were able to read and write, while 0.65 per cent. could merely read, and 0.62 per cent. were unable to read. The proportion who could not read increased slowly with each succeeding quinquennial period of age, until at 50-55 years it stood at 4.04 per cent. At 75 to 80 years the proportion was 7.05, and at 80 and upwards it advanced to 8.07. Similarly, the proportion of persons who could read only increased from 0.65 at 10-15 years to 3.66 at the period 50-55 years, and again to 9.74 and upwards. The better education of the people at the earlier stages is thus exhibited.

Further evidences of improved education will be found in the portion of his work relating to marriages, where it is shown that the proportion of persons in every thousand married, who signed by mark, has fallen very greatly since 1881. The figures for the sexes in the year 1881 were 32.04 males, and 57.04 females, against 6.19 males and 7.02 females in 1895.

For the position of teacher in a public school in New Zealand, at a salary of £60 a year, there were 14 female applicants, 10 of whom held the degree of M.A., and the other four that of B.A.

The number of children, 5-15 years of age, in New Zealand, was estimated as on 31st December, 1902, at 178,875. The number of children, 7-13 years of age (compulsory school age), was estimated as on 31st December, 1902, at 124,986. The attendance at schools, public and private, during the fourth quarter of 1902, was European 150,332, Maoris and half-castes 5,573. If children spend their useful years of child life at school, they can render little or no remunerative service to their parents.

Neither boys or girls can earn anything till over the age of 14 years. Our laws prohibit child labour.

In New Zealand, children, therefore, while they remain at home, are a continual drain on the resources of the bread-winner. More is expected from parents than in many other countries.

At our public schools children are expected to be well clad; and it is quite the exception, even in the poorest localities of our large cities, to see children attending school with bare feet.

During child-life, nothing is returned to the parent to compensate for the outlay upon the rearing and educating of children.

If a boy, by reason of a good education, soon, say, at from 14-18 years, is enabled to earn a few shillings weekly, it is very readily absorbed in keeping him dressed equally well with other boys at the same office or work.

An investment in children is, therefore, from a pecuniary point of view, a failure. There are, perhaps, two exceptions in New Zealand—in dairy farming in Taranaki, where the children milk outside school hours; and in the hop districts of Nelson, where, during the season, all the children in a family become hop-pickers, and a big cheque is netted when the family is a large one.

Quite apart from considerations of self, parents declare that the fewer children they have, the better they can clothe and educate them; and they prefer to "do well" for two or three, than to "drag up" twice or three times as many in rags and ignorance.

Clothing is dear in New Zealand. The following is a labourer's account of his expenditure. He is an industrious man, and his wife is a thrifty Glasgow woman. It is drawn very fine. No. 7 is less than he would have to pay in the city by two or three shillings a week for a house of similar size. No. 9 is rather higher than is usual with Benefit Societies, which average about sixteen shillings a quarter.

WEEKLY EXPENSES OF FAMILY COMPRISING FIVE CHILDREN AND PARENTS.

	Per Week.		
	£	s	d.
1. Groceries and milk	0	15	0
2. Coal and light	0	4	0
3. Butcher	0	4	0
4. Baker	0	4	0
5. Boots, with repairing	0	2	6
6. Clothing and underclothing	0	5	0
7. Rent in suburbs	0	10	0
8. Sundries	0	2	0
9. Benefit Society	0	2	0
Weekly total	£2	8	6

Most young people make a good start in New Zealand. Even men-servants and maid-servants want for nothing. They dress well, they go to the theatres and music-halls, they have numerous holidays, and enjoy them by excursions on land or sea. It is when they marry, and mouths come crying to be filled, that they become poor, and the struggle of life begins.

In our Colony, there is no more prevalent or ingrained idea in the minds of our people than that large families are a cause of poverty.

A high birth-rate in a family certainly is a cause of poverty. Many children do not enable a father to earn higher wages, nor do they enable a mother to render the bread-winner more assistance; while in New Zealand, especially, compulsory education and the inhibition of child-labour prevent indigent parents from procuring the slight help that robust boys and girls of 10 years of age, or so, are often able to supply.

These considerations go far to explain the desire on the part of married couples to limit offspring; and, if there were no means at their disposal of limiting the number of children born to them, a great decline in the marriage-rate would be the inevitable result of the existing conditions of life, and the prevalent ideas of the people.

Hopeless poverty appears to be a cause of a high birth-rate, and this seems to be due to the complete abandonment by the hopelessly poor of all hope of attaining comfort and success.

Marriage between two who are hopelessly poor is extremely rare with us. Each is able to provide for his or herself at least, and in all probability the husband is able to provide comfortably for both.

If he is not, the wife can work, and their joint earnings will keep them from want. But, if one of the partners has not only to give herself up to child-bearing, and thus cease to earn, but also bring another into the home that will monopolise all her time, attention, and energy, and a good deal of its father's earnings, how will they fare?

If a man's wages has to be divided between two, then between three, then four, six, eight, ten, while all the time that wages is not increasing, have we not a direct cause of poverty, and, moreover, is not that cause first in time and importance?

Later on in the history of the family their poverty will become a cause of an increase in the children born to them. At first they may struggle to prevent an increase, but, when they are in the depths of hopeless poverty, they will abandon themselves to despair.

Could they have had born to them only one, or two, or three, during their early married life, they might not only have escaped want, but later in life may have had others born to them, without either their little ones or themselves feeling the pinch of poverty.

It must be remembered in this connection that fecundity and sexual activity are not convertible terms.

It is certainly not true to say that the greater the fecundity of the people the stronger their sexual instinct, or the greater the sexual exercise.

A high fecundity does not depend on an inordinate sexual activity.

Fecundity depends on the child-bearing capacity of each female, and a sexual union at an appropriate time once in two years between puberty and the catamenia is compatible with the highest possible fecundity.

It would be quite illogical, and inconsistent with physiological facts, to aver that, were the poor less given to indulge the pleasures of sense, their fecundity would be modified in an appreciable degree.

CHAPTER VI.

ETHICS OF PREVENTION.

"So God created man in his own image, in the image of God created He him, male and female created He them, and God blessed them and God said unto them, 'Be fruitful and multiply, and replenish the earth.'"—(Genesis i., 27-28). This commandment was repeated to Noah and his sons.

Whether Moses was recording the voice of God, or interpreting a physiological law is immaterial to this aspect of a great social question. The fact remains that in obedience to a great law of life, all living things are fruitful and multiply and replenish the earth, and multiplication in a state of nature is limited only by space and food.

In a state of nature, reproduction is automatic, and only in this state is this physiological law, or this divine command obeyed.

The reason of man intervenes, and interprets, and modifies this law.

A community of men becomes a social organism, calls itself a State, and limits the law of reproduction. It decrees that the sexes shall, if they pair, isolate themselves in pairs, and live in pairs whether inclined to so live or not.

If the State has a right so to interpret and limit the law of reproduction, a principle in human affairs is established, and its decree that individuals shall not mate before a certain age, or not mate at all, is only a further application of the same principle. By the law of reproduction a strong instinct, second only in force and universality to the law of self-preservation, is planted in the sexes, and upon a blind obedience to this force, the continuity of the race depends.

The tendency in the races of history has been to over-population, or to a population beyond the food supply, and there is probably no race known to history that did not at some one period of its rise or fall suffer from over-population.

States have mostly been concerned, therefore, with restraining or inhibiting the natural reproductive instinct of their subjects through marriage laws which protect the State, by fixing paternal responsibility. There were strong reasons why a State should not be over-populated, and only one reason why it should not be under-populated. That one reason was the danger of annihilation from invasion.

Sparta was said to have suffered thus, because of under-population, and passed a law encouraging large families. Alexander encouraged his soldiers to intermarry with the women of conquered races, in order to diminish racial differences and antagonism, and Augustus framed laws for the discouragement of celibacy, but no law has ever been passed decreeing that individuals must mate, or if they do mate that they shall procreate.

Malthus, the great and good philanthropist of Harleybury, a great moralist and Christian clergyman, urged that it was people's duty not to mate and procreate until they had reasonable hope of being able easily to rear, support, and educate the normal family of four, and, if that were impossible, not to mate at all. As a Christian clergyman, Malthus did not interpret the Divine command apart from the consequences of its literal acceptance.

"Be fruitful," meant to Malthus reproduce your kind,—that implied not only bringing babies into the world, but rearing them up to healthy, robust, and prosperous manhood, with every prospect of continuing the process.

"Multiply and replenish the earth" as a command to Noah, meant in the mind of the Rector of Harleybury, "People the earth with men after your own image."

Very little care would be required in Noah's time, with his fine alluvial flats, and sparse population, but in Malthus's time the command could not be fully carried out without labour, self-development, and "moral restraint."

The physiological law is simple and blind, taking no cognisance of the consequences, or the quality of the offspring produced. The divine command is complex. It embodies the reproductive instinct, but restrains and guides it in view of ultimate consequences.

So much for the views and teaching of Malthus. To him no ethical standard was violated in preventing offspring by protracted continence, or lifelong celibacy, provided the motive was the inability so to provide for a family as to require no aid from the state. And it is difficult to escape this conclusion. There is no ethical, Christian, or social law, that directs a man or woman to procreate their kind if they cannot, or have reasonable grounds to think they cannot, support their offspring without aid from others.

There can be, therefore, no just law that decrees that men or women shall marry under such circumstances. In fact most philanthropists think they violate a social and ethical law if they do marry.

But, if with Paul, they resolve that it is better to marry than to burn, is there any law that can or should prevent them selecting the occasions of their union, with a view to limiting fertility.

Abstention is the voluntary hindrance of a desire, when that desire is strongest in both sexes; and as such it limits happiness, and is in consequence an evil *per se*. A motive that will control this desire must be a strong one; such a motive is not necessarily bad. It may be good or evil.

There can be no essential ethical difference between constant continence, prior to marriage, and intermittent continence subsequent to marriage, both practices having a similar motive.

If post nuptial restraint with a view to limiting offspring is wrong, restraint from marriage with the same motive is wrong.

If delayed marriage in the interest of the individual and the State is right, marriage with intermittent restraint is in the same interest, and can as easily be defended.

The ethics of prevention by restraint must be judged by its consequences. If unrestrained procreation will place children in a home where the food and comfort are adequate to their healthful support and development, then procreation is good,—good for the individual, society, and the State.

If the conditions necessary to this healthful support and development, can by individual or State effort be provided for all children born, it is the duty of the individual and of the State to make that effort.

All persons of fair education and good intelligence know what those conditions are, and if they procreate regardless of their absence, that procreation is an evil, and prevention by restraint is the contrary virtue.

It is not suggested, however, that all those who prevent, without or within the marriage bond, do so from this worthy motive, nor is it suggested that all those who prevent are not extravagant in their demand for luxurious conditions for themselves and for their children.

Many require not merely the conditions necessary to the healthful development of each and every child they may bear, but they demand that child-bearing shall not entail hardships nor the prospect of hardships, shall not involve the surrender of any comfort or luxury, nor the prospect of any such surrender.

Whatever doubt may exist in the minds of moralists and philanthropists as to the ethics of prevention in the face of poverty, there can be no doubt that prevention by those able to bear and educate healthy offspring, without hardship, is a pernicious vice degrading to the individual, and a crime against society and the State.

Aristotle called this vice "oliganthropy." Amongst the ancients it was associated with self-indulgence, luxury, and ease. It was the result of self-indulgence, but it was the cause of mental and moral anæmia, and racial decay.

So far in this chapter prevention has been dealt with only in so far as it is brought about by ante-nuptial and post-nuptial restraint. Artificial checks were first brought prominently before the notice of the British Public under the garb of social virtue, about the year 1877 by Mrs. Annie Besant and Mr. Charles Bradlaugh.

These checks to conception, though they are very largely used, can hardly be defended on physiological grounds. Every interference with a natural process must be attended, to some extent at least, with physical injury. There is not much evidence that the injury is great, but in so far as an interference is unnatural, it is unhealthy, and there is much evidence to show that many of the checks advocated and used, are not only harmful but are quite useless for the purpose for which they are sold.

It will be conceded by most, no doubt, that with those capable of bearing healthy children, and those unable to rear healthy ones when born, prevention by restraint, ante-nuptial or post nuptial, is a social virtue, while prevention under all other circumstances is a social vice.

Happiness has been defined as the surplus of pleasure over pain. What constitutes pleasure and what pain varies in the different stages of racial and individual development. In civilized man we have the pleasures of mind supplementing and in some cases replacing the pleasures of sense. We talk, therefore, of the higher pleasures—the pleasures of knowledge and learning, of wider sympathies and love, of the contemplation of extended prosperity and concord, of hope for international fraternity and peace, and for a life beyond the grave. Happiness to the highly civilized will consist, therefore, of the surplus of these pleasures over the pains of their negation.

Self-preservation is the basal law of life, and to preserve one's-self in happiness, the completest preservation, for happiness promotes health, and health longevity.

The first law of living nature then is to preserve life and the enjoyment of it, and the pleasures sought, to increase the sum of happiness will depend on the sentiments and emotions, *i.e.*, on the faculties of mind that education and experience have developed, in the race, or in the individual.

My first thought is for myself, and my duty is to increase the sum of my happiness. But the mental state we call happiness is relative to the presence or absence of this state in others. Even amongst the lower animals, misery and distress in one of the flock militate against the happiness of the others. In a highly developed man true happiness is impossible in the presence of pain and misery in others and *vice versa*; happiness is contagious and flows to us from the joy of others. If the happiness of others then is so essential to my own happiness, I am fulfilling the first law of life and ministering to my own preservation in health and happiness by using my best endeavours to promote this state in others. My material comfort too depends largely on the labour, and love, and the contribution of others in the complex industrial system and division of labour of the higher civilisations. Not only my happiness and health but my very existence depends on the good-will and toil of others. Thus from a purely egoistic standpoint, my first duty to myself is to increase the happiness in others, and, therefore, my first duty to myself becomes my highest duty to society.

My duty to my child is comprehended in my duty to society, *i.e.*, to others. My duty to others is to increase the sum of the happiness of others, and bringing healthy children into the world not only creates beings capable of experiencing and enjoying pleasures, but adds to the sum of social happiness, by increasing the number of social units capable of rendering service to others.

The next great law of life is the law of race preservation. This law comprises the instinct to reproduction and the instinct of parental love. The first and chief function of these instincts in the animal economy is the perpetuation of the race. The preservation of self implies and comprehends the preservation of the race.

My first duty to myself is to preserve myself in health and happiness; but this is best fulfilled and realized in labouring for the health and happiness of others. If this be the universal law, I also am the recipient of others' care, therefore probably better tended and preserved. I save my life by losing it in others.

My second duty, though nominally to Society, is in reality to myself, and it is to preserve myself by preserving the race to which I belong.

Self-preservation therefore, is the first law of life, race preservation the second or subsidiary law.

To fulfil this second law, nature has placed on every normal healthy man and woman the sacred duty of reproducing their kind. Reproduction as a physiological process promotes, both directly and indirectly, the health, happiness and longevity of healthy men and women.

Statistics confirm the popular opinion "that the length of life, to the enjoyment of which a married person may look forward, is greater than that of the unmarried, both male and female at the same age."—(Coghlan).

It is a familiar observation that the mothers of large families of ten and even twice that number are not less healthy nor shorter lived because of the children they have borne. Pregnancy is a stimulus to vitality. Because another life has to be supported, all the vital powers are invigorated and rise to the occasion—the circulation increases, the heart enlarges in response to the extra work, and the assimilative powers of the body are greatly accelerated. During lactation also, the same extra vital work done is a stimulus to a physiological activity which is favourable to health and longevity. The expectancy of life in women is greater than in men all through life, the difference during the child-bearing period of life being about 2.2 years in favour of women.

Statistics and physicians from their observation agree in this, that the bearing of children by normal women, so far from being injurious to health, is as healthful, stimulating, and invigorating a function as the blooming of a flower, or the shedding of fruit, and a mother is no worse for the experience of maternity than is the plant or the tree for the fruit it bears.

The supreme law of society is the law of race-preservation, and the infraction of this law is a social crime. One's duty to society is a higher duty than to one's-self, but the lower duty comes first in our present stage of racial evolution. Instinct prompts to the one, reason—a higher and later, but less respected, faculty—prompts to the other.

But it can be shown that from an egoistic standpoint my duty to the State in this regard is my highest duty to myself.

The parental sacrifice necessary in rearing the normal number of children is infinitesimal compared with the parental advantage.

Parental love is a passion as well as an instinct in normal men and women, and the full play of this passion in its natural state is productive of the greatest happiness.

Vice may restrain, replace, or smother it, but nothing else can damage or adulterate this powerful passion in the human heart.

Low level selfishness, love of low level luxury, diseased imaginings, and unreasonable dreads and fears, are some of the forms of vice that smother this noble passion.

The pursuit of happiness and the higher forms of selfishness would naturally point to parentage.

The ectasy of parental love, the sweet response from little ones that rises as the fragrance of lovely flowers, self-realization in the comfort and joy of family life, the parental pride in the contemplation of effulgent youth, the sympathetic partnership in success, the repose of old age surrounded by filial manhood and womanhood, all go to make a surplus of pleasure over pain. that no other way of life can possibly supply.

What is the alternative?

To miss all this and live a barren life and a loveless old age. Perhaps to bear a child, that, for the need of the educative, elevating companionship of family mates is consumed by self,inheriting that vicious selfishness, which he by his birth defeated, and finding all the forces of nature focussed on his defect, like a pack of hounds that turn and rend an injured mate.

Or a family of one, after years of parental care and love, education and expense, dies or turns a rake, and the canker of remorse takes his place in the broken hearts.

Nature's laws are not broken with impunity—as a great Physician has said, "She never forgives and never forgets."

Self-preservation and race-preservation together constitute the law of life, just as Conservation of Matter and Conservation of Energy constitute the Law of Substance in Haeckels Monistic Philosophy, and the severest altruism will permit man to follow his highest self-interest in obedience to these laws. It is only a perverted and vicious self-interest that would tempt him to infraction.

That the vice of oliganthropy is growing amongst normal and healthy people is a painful and startling fact. In New Zealand the prevailing belief is that a number of children adds to the cares and responsibilities of life more than they add to its joys and pleasures, and many have come to think with John Stuart Mill, that a large family should be looked on with the same contempt as drunkenness.

CHAPTER VII.

WHO PREVENT.

It has been sufficiently demonstrated in preceding chapters, that the birth-rate has been, and is still rapidly declining. It has been sought to prove that this decline is chiefly due to voluntary means taken by married people to limit their families, and that the desire for this limitation is the result of our social system.

The important question now arises. Is the desire uniform through all classes of Society, and is the practice of prevention uniform through all classes?

In other words, is the decline in the birth-rate due to prevention in one class more than in another, and if so which?

Experience and statistics force us to the startling conclusion, that the birth-rate is declining amongst the best classes of citizens, and remains undisturbed amongst the worst.

Now the first-class responsible for the decline includes those who do not marry, and those who marry late. The Michigan vital statistics for 1894 show that the mean number of children to each marriage at the age of 15-19 years is 6.75, at the age of 20-25 years it is 5.32, a difference of 1.44 in favour of delayed marriage for a period of five years.

In New Zealand the marriage rate has gone up from 5.97 per thousand persons living in 1888 to 7.67 in 1900.

This class includes clerks with an income of £100 and under,—a large number with £150, and all misogynists with higher incomes.

It includes labourers with £75 a year and under, and many who receive £100.

Their motives for avoiding marriage are mostly prudential.

Those who abstain from marriage for prudential reasons are as a rule good citizens. They are workers who realise their responsibilities in life, and shrink from undertaking duties which they feel they cannot adequately perform. By far the largest class who practice prevention, consists of those who marry, and have one or two children, and limit their families to that number, for prudential, health, or selfish reasons.

These too are as a rule good citizens, and there are two qualities that so distinguish them. First, their prudence; they have no wish to burden the State with the care or support of their children. Their fixed determination is to support and educate them themselves, and they set themselves to the work with thriftiness and forethought.

In order to do this, however, it is essential that the family is limited to one, two, or three, as the case may be, and before it is too late, preventive measures are resorted to.

The second quality that distinguishes them as good citizens is their self-control. Every preventive measure in normal individuals implies a certain amount of self-restraint, and in proportion as prudential motives are strong is the self-imposed restraint easy and effective.

The existence of these two qualities, prudence and self-control, is a very important factor in human character, and upon their presence and prevalence in its units depend the progress and stability of society. But the birth-rate varies in an inverse ratio with these qualities. In those communities or sections of communities, where these qualities are conspicuous, will the birth-rate be correspondingly low.

There is another class of people that has strong desires to keep free from the cares and expense of a large family. These are, too, good citizens and belong to good stock. They are those possessed of ambition to rise socially, politically, or financially, and they are a numerous body in New Zealand.

They are quite able to support and educate a fairly large family, but as children are hindrances, and increase the anxieties, the responsibilities and the expense, they must be limited to one or two.

There is still another class that consists of the purely selfish and luxurious members of society, who find children a bother, who have to sacrifice some of the pleasures of life in order to rear them.

Now all those who prevent have some rational ground for prevention, and at least are possessed of sufficient self-control to give effect to their wish. They include the best citizens and the best stock, and from them would issue, if the reproductive faculty were unrestrained, the best progeny.

One grave aspect of this limitation is that, as a rule, the family is limited after the first one or two are born. The small families, say of two, are born when the parents are both young, and carefully compiled statistics prove that these are not the best offspring a couple can produce. Those born first in wedlock, are shorter and not so well developed as those born later in married life, when parents are more matured.

If it is substantially true, that the decline in the birth-rate is due to voluntary prevention, and that prevention implies prudence and self-control, it is safe to conclude that those in whom these qualities are absent or least conspicuous, will be the most prolific.

But those in whom these qualities are absent or least conspicuous are our worst citizens, and, therefore, our worst citizens are the most prolific. Observation and statistics lead to the same conclusion.

Amongst the very poor in crowded localities, the passion for marriage early asserts itself.

Its natural enemies are prudence and a consciousness of responsibility, and these suggest restraint. But prudence and restraint are not the common attributes of the very poor. Poverty makes people reckless, they live from hour to hour as the lower animals do. They satisfy their desires as they arise, whether it be the desire for food or the desire of sex.

The very poor includes amongst its numbers, the drunkard, the criminal, the professional pauper, and the physically and mentally defective.

The drunkard is not distinguished by his prudence, nor by his self-restraint. In fact the alcohol which he imbibes paralyses what self-control he has, and excites through an increased circulation in his lower brain-centres an unnatural sexual desire. What hope is there of the drunkard curtailing his family by self-restraint?

Dr. Billings says, (Forum, June 1893) "So far as we have data with regard to the use of intoxicating liquors, fertility seems greatest in those countries and amongst those classes where they are most freely used."

Neither is the criminal blessed with the important attributes of prudence and self-control. They are conspicuous by their absence in him.

In all defectives, in epileptics, idiots, the physical deformed, the insane, and the criminal, the prudence and self-restraint necessary to the limitation of families is either partially or entirely absent.

To the poor in crowded localities, with limited room-space and insanitary surroundings, effective self-restraint is more difficult than in any other class of society.

In all defectives the sexual instinct is as strong, if not stronger, than in the normal, and they have not that interest in life, and regard for the future that suggest restraint, nor have they the power to practise it though prudence were to guide them.

The higher checks to population, as they exist among the better classes of people, do not obtain amongst the defectives taken as a class.

Vice and misery are more active checks amongst the very poor, and abortion is practised to a very considerable extent, but the appalling fact remains, that the birth-rate of the unfit goes on undisturbed, while the introduction of higher checks amongst the normal classes has led to a marked decline, more marked than at first sight appears. The worst feature of the problem, however, is not so much the disproportion in the numbers born to the normal and the abnormal respectively, but the fact that the defectives propagate their kind.

The defectives, whose existence and whose liberty constitute the greatest danger to the State, are the intermittent inhabitants of our lunatic asylums, prisons, and reformatories.

There is one defect common to all these, and that is defective inhibition.

All human activity is the result of two forces, motor impulses tending to action, and inhibition tending to inertia.

The lower animals have strong motor impulses constantly exploding and expressing themselves in great activity, offensive, defensive, self-preservative, and procreative, being restrained only by the inhibitive forces of their conditions and environment.

Children have strong motor impulses, which are at first little controlled. Inhibition is a late development and is largely a result of education.

If the motor impulses remain strong, or become stronger in the presence of development with exercise, while inhibition remains weak, we have a criminal.

Inhibition is the function performed by the highest and last-formed brain-cells. These brain cells may be undeveloped either from want of exercise, that is, education, or from hereditary weakness, or, having been developed may have undergone degeneration, under the influence of alcohol, or from hereditary or acquired disease.

Motor impulses, as the springs of action, are common to all animals. In the lower animals inhibition is external, and never internal or subjective. In man it may be internal or external.

It is internal or subjective in those whose higher brain centres are well developed and normal. Their auto-inhibition is such that all their motor impulses are controlled and directed in the best interests of society.

It is external only in those whose higher brain centres are either undeveloped or diseased. These constitute the criminal classes. Their motor impulses are unrestrained. They offer a low or reduced resistance to temptation.

Weak or absent resistance in the face of a normal motor impulse whose expression injuriously affects another, is crime, and a criminal is one whose power of resistance to motor impulses has been reduced by disease, hereditary or acquired, or is absent through arrested development.

A confirmed criminal is one in whom the frequent recurrence of an unrestrained impulse injurious to others has induced habit.

Auto-inhibition is defective or absent, and society must in her own interest provide external restraint, and this we call law.

Criminals are, therefore, mental defectives, and may be defined for sociological purposes as those in whom legal punishment for the second time, for the same offence, has failed to act as a deterrent.

M. Boies, in "Prisoners and Paupers," says that conviction for the third time for an offence, is proof of hereditary criminal taint.

The existence of motor impulses in the human animal is normal. They vary in strength and force. We cannot eradicate, we can only control them.

They may become less assertive under the constant control of a highly cultivated inhibition, but it is only in this way that they can be affected at all. They may be controlled, either by the individual himself or by the State. Our reformatories are peopled by young persons whose distinguishing characteristic is that inhibition is undeveloped or defective. This defect may be due to want of education, but it is more often hereditary.

Two things only can be done for them. This faculty of inhibition can be trained by education, or external restraint can be provided by law.

But the distinguishing characteristic of all defectives, within or without our public institutions, is defective inhibition,—they are unable to control the spontaneous impulses that continually arise, and which may indeed be normal.

Impulses may be abnormal from hereditary predisposition, as *e.g.* the impulse to drink, but only through strengthening inhibition can these impulses be controlled,—their existence must be accepted.

But whether the defect is an abnormal impulse, or a normal impulse abnormally strong, or an abnormally weak or defective inhibition, the condition is hereditary, and such defectives propagate their kind.

It has been shown that they are more fertile than any other classes because of the very defect that makes them a danger to society.

The defective restraint that allows them to commit offences against person and property, also allows their procreative impulse unrestrained activity.

Defectives, therefore, are not only fertile, but they propagate their kind, and a few examples will serve to show to some extent the fertility, and to an enormous extent the hereditary tendencies, of the unfit.

CHAPTER VIII.

THE MULTIPLICATION OF THE FIT IN RELATION TO THE STATE.

The multiplication of the fit is of the first importance to the State. It supplies competent producers and courageous defenders, and the more of these, consistent with space and food (using these terms in their fullest significance), the better off the State.

If healthy happy citizens are the State's ideal, then limitation of population well within the space and food will be encouraged. If national wealth and prosperity in its material aspect are the State's ideal, the harder the population presses on the means of subsistence the sooner will that ideal be realised. For it cannot be denied, that the greater the stress and hardship in life, the more strenuous the effort put forth to obtain a foothold. The greater the competition the keener the effort, and the higher the accomplishment; while to ensure an adequate supply of labour in time of great demand there must always be a surplus.

The waste of life must always be greater; but what of that! National wealth is the ideal—the maximum amount of production. Child labour, and women labour, are called in to fill the national granaries, though misery and death attend the process.

If this be the ideal of the State, life is of less value than the product of labour, for it can be more easily and readily replaced.

But the ideal of the perfect state is not wealth but the robust happiness of its members.

The happiness of its members is best promoted by the maximum increase in its numbers, consistent with ample space and food. With ample space and food multiplication works automatically, being kept up to the limit of space and food by the procreative instinct.

If it can be shown that multiplication is not sufficiently stimulated by this instinct, then it must be concluded that, *in the minds of the citizens* the space and food are not ample.

In New Zealand the procreative impulse does not keep multiplication at an equal pace with the apparent supply of food and space, and this is due, as has been shown, to the fact that our citizens are not satisfied that the supply *is* ample.

They have come to enlarge the definition of "food," and this term now includes luxuries easily obtainable for themselves and their families.

But the luxuries of life and living can only be easily obtained when individual effort to obtain them is unhampered. Every burden which a man has to bear (only the best are here referred to,—the fit members of the State) limits his power to provide for himself, and any he may bring into the world.

If the State decrees that a citizen shall support himself, his mate, and his progeny, well and good,—if he has no other burden to bear, no other responsibility, he knows exactly where he is and what he has to do, and directs his energies and controls his impulses, and enlarges his desires to suit his tastes and purposes.

But if the State decrees that a citizen shall not only support all for whose existence he is responsible, but also all those unable to support themselves, born into the world in increasing numbers as congenital defectives, and manufactured in the world by legalised drinking saloons, and by pauperising charitable aid and benevolent institutions, then our self-respecting right-respecting citizen must decide whether he will forego the luxury and ease that he may enjoy, and rear the normal family, or curtail his own progeny, and support the army of defectives thrown upon society by the State-encouraged fertility of the unfit.

It has already been shown, that in this colony the best fit to multiply are ceasing to do so, because of a desire to attain a social and financial stability that will protect them and their dependents from want or the prospect of want. There is every reason to believe, that when this stability is assured the normal family soon follows.

The love of luxurious idleness and a passion for excitement, which were typical of the voluntarily barren women of ancient Rome, have little place with us, as a cause of limited nativity.

Men and women reason out, that they cannot bear all the burdens that the State imposes upon them, support an increasing army of paupers, and lunatics and defectives, and non-producers, and that luxuriously, and at the same time incur the additional burden of rearing a large family.

Let us examine these burdens, and see if the complaint of our best stock is justified.

The amount raised by taxation in New Zealand (including local rates) during the year 1902-03, amounted per head of population (excluding Maories) to £5 4s. 7d. The bread-winners in New Zealand number according to official returns, 340,230, and the total rates and taxes collected for the year 1902-03 amounted to £4,174,787 or £12 5s. 4d. for each bread-winner for the year.

On March 31st, 1901 (the last census date) there were 23.01 persons per thousand of population over 15 years of age, unable to work from sickness, accident and infirmity. Of these 12.72 were due to sickness and accident, and 10.29 to "specified infirmities."

The proportion of those suffering from sickness and accident in 1874 was 12.64 per 1000 over 15 years, practically the same as for 1901, while disability from "specified infirmities" (lunacy, idiocy, epilepsy, deformity, etc.)—degeneracies strongly hereditary—rose rapidly from 5.32 in 1874 to 10.29 in 1901, or taking the total sickness and infirmity, from 17.96 in 1874 to 23.01 in 1901.

On the last census date there were 340,230 bread-winners, and 12,747 persons suffering from sickness, accident, and infirmity, or 26 fit to work and earn for every one unfit.

The cost to the Colony per year of—

		£
1.	Hospitals, year ended 31st March, 1903	138,027
2.	Charitable Aid (expended by boards), year ended 31st March, 1903	93,158
3.	Lunatic Asylums, year ended 31st Dec, 1902 (gross)	85,238

	Lunatic Asylums, year ended 31st Dec, 1902 (nett)	64,688
4.	Industrial Schools, year ended 31st Dec, 1902	
	Government Industrial Schools for neglected and criminal children	21,708
	Government Expenditure on Private Denominational Industrial Schools	2,526
5.	Police Force, year ended 31st March, 1903	123,804
6.	Prisons, year ended 31st March, 1903	32,070
7.	Criminal Courts (Criminal Prosecutions), year ended 31st March, 1903	16,813
8.	Old Age Pensions (pensions only for persons over 65 years of age, who have been 25 years in the Colony, and who make a declaration of poverty, including departmental expenses)	212,962

A total of £705,756. This constitutes the burden due to defectives and defects in others, a handful of workers have to bear in a sparse population of 800,000 souls in one of the finest countries on which the sun of heaven ever shone.

The burden which the fit have to bear has often been referred to by Dr. MacGregor, who states in one of his reports, "Wives and husbands, parents of bastards, all alike are encouraged by lavish charity (falsely so called) to entirely shirk their responsibilities in the well grounded assurance that public money will be forth-coming to keep them and their families in quite as comfortable position as their hardworking and independent neighbours."

The state can not decree that men shall marry, or that women shall marry, or that women shall procreate. All it can do is to discover why its subjects are not fertile, and remove the causes so far as it is possible.

As people become educated they become conscious of their limitations, and endeavour to break through them and better their conditions.

The more difficult this process is, the less likely will men and women be to incur the burden of a large family. The more the conditions of existence are improved, the more completely is each man's wish realized, and the more readily will he undertake the responsibilities of a family.

If the State can and will lighten the burden of taxation and modify the strain and stress of life, it will indirectly encourage procreation.

No direct encouragement is possible. It was tried and it failed in Sparta, it was tried by Augustus and it failed in Rome, it must fail everywhere, for the most willing and the most ready to respond to any provision made to encourage increase, are the unfit, and it is the fertility of the unfit that is the very evil that has to be attacked.

It is the fertility of the unfit that makes the burden of the fit, and a tax on bachelors, or a bonus on families, would be responded to by the least fit, long before it affected those whose response was anticipated, and the problem sought to be solved would only be aggravated thereby.

No encouragement whatever can the State afford to give to the natural increase of population till it has successfully grappled with the propagation of defectives.

The burden of life would be lessened by nearly one-third if the fertility of defectives could be stopped.

The State would have to support only those who acquired defects, the scars of service more honourable than wealth, in their efforts to support themselves and families, and these would be few indeed, if inherited tendencies could be eliminated or reduced to a minimum.

It is the purpose of this work to attempt to describe a method that will help to bring about this end.

CHAPTER IX.

The Multiplication of the Unfit in Relation to the State.

THE humanitarian spirit, born 1900 years ago, effectually checked all inhuman practices for disposal of the unfit. Christ is the Author of this spirit. The noisy triumph of His persecutors had scarcely died away before His conception of the sanctity of human life found expression in the mission of those Roman maidens who in His name devoted their lives to collecting exposed infants from the environs of their city—that they might rear and educate them and bring them to the Church.

Not only has it done this, but it has taught society that its first and highest duty is to its weaker brethren, who constitute the unfit. All our modern institutions are based on this sentiment, and what is the result? Weaklings are born into the world and the weaker they are the more carefully are they tended and nursed. The law of the struggle for existence, *i.e.,* the law of Justice is suspended or modified, and the unfit are allowed to live, or at least allowed to live a little longer, long enough indeed to propagate their kind.

Hospitals and Homes and Charitable institutions all combine their energies, and direct their efforts to nurture those whom the laws of nature decree should die.

Sympathy and not indignation is aroused when a defective is born, and the result of all the effort which that sympathy evokes is that the little weakling and thousands such are safely led and tended all the way to the child-bearing period of life, only to repeat their history, in others.

Not only do defects "run in families," but they run in groups, and a physical defect such as club-foot, cleft palate, or any arrested development, is apt to be associated with some mental defect, and it is the mental more than the physical defects of individuals that prevent them being self-supporting helpful members of society.

In the "North American Review" for August, 1903, Sir John Gorst declares that:—

"The condition of disease, debility, and defective sight and hearing, in the public elementary schools in poorer districts, is appalling. The research of a recent Royal Commission has disclosed that of the children in the public schools of Edinburgh, 70 per cent, are suffering from disease of some kind, more than half from defective vision, nearly half from defective hearing, and 30 per cent, from starvation. The physical deterioration of the recruits who offer themselves for the army is a subject of increasing concern. There are grounds for at least suspecting a growing degeneracy of the population of the United Kingdom, particularly in the great towns."

The following table gives the charges before Magistrates in our Courts:—

Year.	Proportion per thousand of mean population.
1894	24.76
1897	26.87
1898	29.42
1899	29.48
1900	31.54
1901	33.20
1902	35.19

Now who are the unfit? Are they more fertile than the fit? and do they propagate their kind?

The following defects constitute their victims members of that great class of degenerates who are unfit to procreate healthy normal offspring. Many of these conditions are partly congenital and partly acquired, but in the majority of defectives a transmitted taint is present.

- I. Congenital defects:—

1. Idiocy.
2. Imbecility.
3. Criminal Taint.
4. Insanity.

5. Inebriate Taint.
6. Pauperism.
7. Deaf Mutism.
8. Epilepsy.

- II. Acquired defects:—

1. Crime.
2. Insanity.
3. Epilepsy.
4. Inebrity.
5. Confirmed Pauperism.

With the exception of the very young and the very old, all members of society, who have to be supported by others, constitute the unfit. Many are supported by friends and relatives, but year by year, it is becoming more noticeable, that the moral guardians of the unfit are shirking their responsibility and handing their defective relatives over to the State and demanding their gratuitous support as a right.

Dr. MacGregor, Inspector of Asylums and Hospitals, N.Z., in his report for 1898, says:—

"As if the State had a vested interest in the degradation of its people, I find that they, as fathers, mothers, brothers, sisters, are responding to our efforts to sap their self-respect by doing their utmost to throw the cost of maintaining their relatives on the ratepayers. I constantly hear the plea urged that as taxpayers and old colonists they have a right to send their relatives to State institutions."

Our social conditions manufacture defectives, and foster their fertility. The strain and stress of modern competition excite an anxiety and nervous tension under which many break down, and much of the insanity that exists to-day is attributable to nervous strain in the struggle of life.

The strong attractive force of one social stratum upon the next below, excites in the latter a nervous tension which predisposes to a breakdown in the face of some adversity.

The passion for ease and luxury, and the dread of poverty tend to overstrain the nervous system, and numberless neurotic defectives fall back upon society, and give themselves up to the propagation of their kind.

Our charitable aid institutions tend largely to swell the numbers of the great unfit.

Dr. MacGregor in one of his valuable and forcible reports upon our charitable aid institutions, says:—

"Our lavish and indiscriminate outdoor relief, whose evils I am tired of recapitulating,—our shameless abuse of the hospital system,—the crowding of our asylums by people in their dotage, kept there because there is no suitable place to send them to, and many of them sent by friends anxious only to be relieved of the duty of supporting and caring for them,—what is it all coming to?"...

"The practical outcome of our overlooking the continued accumulation of degenerates among our people by our fostering of all kinds of weakness will necessarily be, if it continues, that society will itself degenerate. Taxation will increase by leaps and bounds, and the industrious and self-respecting citizens will rebel, especially if taxation is expected to meet all the demands of a legislature that puts our humanitarian idea of justice in the place of charity."

It has already been urged that there is no evidence of any physiological defect in any class of society interfering with fertility. Sexual inhibition, from prudential motives is the real cause in New Zealand.

Sexual inhibition implies well-developed self-control, the very force in which almost all defectives are most deficient, and the absence of which makes them criminals, drunkards and paupers. In almost all defectives too, prudence is conspicuous by its absence.

The only moral force we know of, that has curtailed, or will curtail, the family within the limits of comfortable subsistence, is sexual inhibition with prudence. But this force is absolutely impossible amongst defectives.

It is not only a powerful force among the normal, but with us to-day it is powerfully operative. Amongst the defectives it does not and cannot exist.

Apart from observation and statistics, therefore, it can be shown that the birth-rate amongst the unfit is undisturbed. They marry and are given in marriage, free from all restraint save that of environment, and worst of all they propagate their kind.

Dr. Clouston says "As we watch children grow up we see that some have the sense of right and wrong, the conscience, developed much sooner and much stronger than others; just as some have their eye teeth much sooner than others; and looking at adults, we see that some never have much of this sense developed at all. This is notoriously the case in some of those whose ancestors for several generations have been criminals, insane or drunkards." Again "We know that some of the children of many generations of thieves take to stealing, as a young wild duck among tame ones takes to hiding in holes, and that the children of savage races cannot copy at once our ethics nor our power of controlling our actions. It seems to take many generations to redevelop an atrophied conscience. There is no doubt that an organic lawlessness is transmitted hereditarily."

Mr. W. Bevan Lewis says "It is also notable, that in a large proportion of cases, we find the history of ancestral insanity attached to the grand-parents, or the collateral line of uncles and aunts, significant of a more remote origin for the neurosis. The actual proportion of cases revealing strongly-marked hereditary features (often involving several members of the subject's ancestry), amounts to 36 per cent;" while Mr. Briscoe declares (Journal of Mental Science, Oct. 1896) that 90% of the insane have a heredity of insanity.

The following table from Dr. MacGregor's reports gives an account of two families in New Zealand and their Asylum history.

Number. Name.	Cost per head. Rate £1 Per week.	Total Cost.

	Family of B (Brothers).	£ s. d.	£ s. d.
I.	A.B.	80 2 0	
II.	C.B.	274 4 0	
III.	D.B.	230 2 0	
IV.	E.B.	8 2 0	
V.	F.B.	8 2 0	
		----	600 12 0
	Family of C.		
I.	A.C. (wife)	472 2 0	
II.	B.C. (husband of A.C.)	418 0 0	
III.	D.C. (daughter of A.C.)	834 2 0	
IV.	E.C. (ditto)	1,318 2 0	
V.	F.C. (illegitimate daughter of E.C.)	169 8 0	
VI.	G.C. (husband of F.C.but no blood relation)	5 2 0	
		----	3,216 16 0

			£3,817 8 0

In his report for 1897, the same writer says:—"I know of a 'defective' half-imbecile girl, who has had already five illegitimate children by different fathers, all of whom are now being supported by the Charitable Aid Board, while, of course, the mother is maintained, and encouraged to propagate more;" while in an appendix to a pamphlet on "Some Aspects of the Charitable Aid question," he gives the following history of two defective cases:—

J.A. admitted to Lunatic Asylum, May, 1897.

Three medical men report on her as follows:—"She appears imbecile, but without delusions: natural imbecility, stupid, idiotic expression; baby one month old; age between 30 and 40. Suffering from dementia; lactational."

J.A., husband aged 69; labourer, average earnings 15s. week. He wishes to get admission into some Old Man's Home.

This couple have six children—four girls and one boy. A. aged 12; B. 10; C. 9; D. (boy) 5; and E. 3 years. These children are all in the Industrial School. There is also one baby, born April, 1897; has been put out to nurse by the County Council.

The sister of Mrs. J.A. in Salvation Army Home. There are two brothers, whereabouts not known. The police report on this case that the whole of the relatives of Mrs. J.A. were partly imbecile, always in a helpless condition and state of destitution, and have been for years supported partly by charity of neighbours and help from the Charitable Aid Boards.

J.J., the father, now dead, reported as a "lazy, drunken fellow."

A.J., the mother, "a drunken prostitute" (police report 1886). "Makes a precarious living at nursing" (police report 1897); in destitute circumstances, living with a man known as a thief.

This couple had seven children—six boys and one girl:—

A., committed to Industrial School, 1877; discharged from there 1890; aged 18. Sentenced in 1896 to three years for burglary.

B., committed to Industrial school for larceny in 1883; discharged from there, 1887; aged 17.

C., committed to Industrial School for breaking into and stealing, 1886; aged 16; discharged, 1890.

D., aged 14; E. 9½; and F., 7 years; were sent to Industrial School in 1891 by the Charitable Aid Board, the father being dead and the mother in gaol.

D. was discharged last year, aged 18. F. is in hospital for removal of nasal growth, and defective eyesight. E. was admitted to a lunatic Asylum, September, 1897. Four medical men report on him as follows:—"A case of satyriasis from congenital defect." "His depraved habits result of bad bringing up by his mother." "Probably hereditary." "A case of moral depravity associated with mental deficiency, and cretinism." The youngest of the family, a girl aged 11, is said to be dependent on her mother.

With regard to the hereditary nature of Insanity, John Charles Bucknill and Daniel Hack Tuke, M.D.'s, in "A Manual of Psychological Medicine," says:—

"Certainly, if in ever so small degree there is to be a stamping out of insanity, we must act on the principle, better let the individual suffer than run the risk of bequeathing a legacy of insanity to the next generation.... With regard to males, marriage would no doubt be highly beneficial in many instances, *and if the risk of progeny is not run, may well be encouraged*."

Esquirol, quoted by Bucknill and Tuke, says:—"Of all diseases Insanity is the most hereditary."

Bucknill and Tuke, say:—

"Of marriage it may be said that the celibacy of the insane is the prophylaxis of Insanity in the race, and although a well chosen mate and a happy marriage may sometimes postpone or even prevent the development of insanity in the individual, still no medical man, having regard to the health of the community, or even of that of the family, can possibly feel himself justified in recommending the marriage of any person of either sex in whom the insane diathesis is well marked."

Again "It is thus that the seeds of mental diseases and of moral evils are sown broadcast through the land; and other new defects and diseases are multiplied and varied with imbecilities, and idiocies, and suicidal and other propensities and dispositions, leading to all manner of vice and crime. The marriage of hereditary lunatics is a veritable Pandora's box of physical and moral evil."

The least fit, then, are the most fertile, and the most fertile are subject to the common law of heredity, and the defects are transmitted to their offspring, often accentuated by the intermarriage which their circumstances favour or even necessitate.

But this is not all. The least fit have the worst environment, and in the worst possible surroundings the progeny of the unfit multiply and develop. They are born into conditions, well described by Dr. Alice Vicery, in a paper on "The food supplies of the next generation." "Conditions in which the food, warmth, and clothing which are necessary for the mere maintenance of the functions of the body in their normal state, cannot be obtained; in which men, women, and children are forced to crowd into dens wherein decency is abolished, and the most ordinary conditions of healthful existence are impossible of attainment; in which the pleasures within reach are reduced to bestiality and drunkenness; in which the pains accumulate at compound interest in the shape of starvation, disease, stunted development, and moral degradation in which the prospect of even steady and honest industry is a life of unsuccessful battling with hunger, rounded by a pauper's grave."

What possible hope can there be for the progeny of defectives born with vicious, criminal, drunken or pauper tendencies, into an environment whose whole influence from infancy to maturity tends to accentuate and develop these inherited defects?

In this pitiable stratum of human society, vice and misery, as checks to increase, reign supreme, but as no other check exists, fertility is at its maximum, and keeps close up on the heels of the positive checks.

The State in her humanitarian sympathy, and in New Zealand it is extravagant, puts forth every effort to improve the conditions of its "submerged tenth." Insanitary conditions are improved, the rooms by law enlarged, the air is sweetened, the water is purified, the homes are drained. The delicate and diseased are taken to our hospitals, the deaf and blind to our deaf-mute institutions, the deformed and the fatherless to our orphan homes. And all are carefully nursed as tender precious plants. They are snatched from Nature's clutch and reared as prize stock are reared and kept in clover, till they can propagate their kind.

We feed and clothe the unfit, however unfit, and then encourage their procreation, and as soon as they are matured we foster their fertility.

No want of human sympathy for the poor unfortunates of our race is in these words expressed,—a statement simply of the inevitable consequences of unscientific and anti-social methods of dealing with the degenerate.

No State can afford to shut its eyes to the magnitude of this problem. The procreation of the unfit must be faced and grappled with. And the greater the decline in the birth-rate of our best stock, the more urgent does the solution of the problem become. For is not the proportion of the unfit to the fit yearly increasing!

It has become the most pressing duty of the State, in face of the great change that has so rapidly come over our natural increase, to declare that the procreation of the unfit shall cease, or at least, that it shall be considerably curtailed and placed among the vanishing evils, with a view to its final extinction.

CHAPTER X.

WHAT ANÆSTHETICS AND ANTISEPTICS HAVE MADE POSSIBLE.

For the intelligent mind, which I assume has already been impressed with the importance of such an inquiry, I think I have set forth the salient truths with sufficient clearness, but holding that a recitation of social faults, without a suggestion as to social reforms, is not only useless but mischievous, I shall endeavour to show not only that the situation is not hopeless, but that science and experience have, or will reveal means to the accomplishment of all rationally desired ends, and that it remains only for intelligence to enquire that sentiment may move up to the line so as to harmonise with science, with justice, and with the demands of a growing necessity.

These questions of population are not new. More than two thousand years ago, many of the wisest philosophers of all the centuries meditated deeply upon the tendencies of the population to crowd upon subsistence, and in many ages and many countries, the situation has been discussed with serious forebodings for the future.

In all ages thinking men have regarded war with aversion, yet with peace and domestic prosperity other dangers arose to threaten the progress of the race, and as the passing generations cried out for some remedy for the ever pressing evils, thinking men have been proposing measures somewhat harmonising with the knowledge or the sentiment of the times. Whether we are wiser than our ancestors remains an unsettled question.

The old Greeks faced the problem boldly. There were two dangers in the minds of these ancient philosophers. There was the danger of over-population of good citizens, and there was the danger of increasing the burden good citizens had to bear by the maintenance of defectives. However good the breed, over-population was an economic danger, for, said Aristotle, "The legislator who fixes the amount of property should also fix the number of children, for if the children are too many for the property the law must be broken." And he further declares "As to the exposure and rearing of children, let there be a law that no deformed child shall live"; and the exposure of infants was for years the Grecian method of eliminating the unfit.

A century ago "Parson Malthus" dealt with over-population without regard to the fitness of individuals to survive, and he advised the exercise of moral restraint expressed in delayed marriage, to prevent population pressing on the limits of food, which he maintained it invariably tends to do. After the high souled Malthus, came the Neo-Malthusians, who, although they retained the name perverted the teaching of this great demographist, and some Socialist writers of high repute still advocate the systematic instruction of the poor in Neo-Malthusian practices.

The rising tide of firm conviction in the minds of present day sociologists, that the fertility of the unfit is menacing the stability of the whole social superstructure, is forcing many to advocate more drastic measures for the salvation of the race. Weinhold seriously proposed the annual mutilation of a certain portion of the children of the popular classes. Mr. Henry M. Boies, the most enlightened analyst of the problem of the unfit, in his exhaustive work "Prisoners and Paupers," urges the necessity of effectively controlling the fecundity of the degenerate classes, and he points to surgery, and life-long incarceration as the solution of the problem. Dr. McKim, in an exhaustive work on "Heredity and Human Progress," after declaring that he is profoundly convinced of the inefficiency of the measures which we bring to bear against the weakness and depravity of our race, ventures to plead for the remedy which alone, as he believes, can hold back the advancing tide of disintegration. He states his remedy thus:—"The roll then, of those whom our plan would eliminate, consists of the following classes of individuals coming under the absolute control of the State:—idiots, imbeciles, epileptics, habitual drunkards and insane criminals, the larger number of murderers, nocturnal house-breakers, such criminals whatever their offence as might through their constitutional organization appear very dangerous, and finally, criminals who might be adjudged incorrigible. Each individual of these classes would undergo thorough examination, and only by due process of law would his life be taken from him. The painless extinction of these lives would present no practical difficulty—in carbonic acid gas we have an agent which would instantaneously fulfil the need."

These briefly are some of the remedies which have been advocated and in part applied for the protection of the race from degeneracy. I quote them, not with approval, but merely to show how grave and serious the social outlook is, in the minds of some of the best thinkers and truest philanthropists that have taught mankind. If the fertility of the fit could be kept uniformly at its normal rate in a state of nature, the race would have little to fear, for the tendency to further degeneration and consequent extinction amongst the defective would be sufficient to counteract their disposition to a high fertility. But in all civilized nations, the fertility of the fit is rapidly departing from that normal rate, and Mr. Herbert Spencer declares, with the gloomiest pessimism, that the infertility of the best citizens is the physiological result of their intellectual development. I have already expressed the opinion that prudence and social selfishness, operating through sexual self-restraint on the part of the best citizens of the State, are the cause of their infertility. It is impossible for the State to correct this evil, except by lessening the burden the fit man has to bear; and the elimination of the unfit, by artificial selection, is the surest and most effective way of bringing this about.

We have learned from the immortal Pasteur the true and scientific method of artificial selection of the fit, by the elimination of the unfit. We have already seen that he examined the moth, to find if it were healthy, and rejected its eggs if it were diseased. Medical knowledge of heredity and disease makes it possible to conduct analogous examinations of prospective mothers; and surgery secreted in the ample and luxurious folds of anæsthesia, and protected by its guardian angels antiseptics, makes it possible to prevent the fertilization of human ova with a vicious taint. It is possible to sterilize defective women, and the wives of defective men by an operation of simple ligature, which produces absolutely no change whatever in the subjects of it, beyond rendering this fertilization impossible, for the rest of life. This remedy for the great and growing evil which confronts us to-day is suggested, not to avenge but to protect society, and in profound pity for the classes who are a burden to themselves, as well as to those who have to tend and support them.

The problem of the unfit is not new. The burden of supporting those unable to support themselves has been keenly felt in all ages and among all peoples.

The ancients realized the danger and the burden, but found no difficulty when the stress became acute in enacting that all infants should be examined and the defective despatched.

To come nearer home, Boeltius tells us, that, "in old times when a Scot was affected with any hereditary disease their sons were emasculated, their daughters banished, and if any female affected with such disease were pregnant, she was to be burned alive."

Aristotle declared that "neglect of this subject is a never failing cause of poverty, and poverty is the parent of revolution and crime," and he advocated habitual abortion as one remedy against over-population. The combined wisdom of the Greeks found no better method of keeping population well within the limits of the State's power to support its members than abortion, and the exposure of infants.

Since Aristotle's time abortion has been largely practised by civilized nations. Mutilation and infibulation of females have been practised by savages with the same end in view, while vasectomy, orchotomy, and ovariotomy, have had their avowed advocates in our own time.

The purpose of all these measures was to limit population with little or no distinction as to fitness to survive. The Spartans in ancient times, and many social reformers of to-day have discussed and advocated the artificial limitation of the unfit. The exposure of defective infants was the Spartan method of preserving the physical and mental stature of the race.

The surgical operations on both sexes advocated by some social writers of recent date, have not been received with much favour, and, as a social reform have not been practised. As operations they are grave and serious, profound in their effect upon the individual, and a violation of public sentiment. Anæsthetics and antiseptics have, however, made them possible, and if a surgical operation could be devised, simple and safe in performance, inert in every way but one, and against which there would be no individual or public sentiment, its application as a social reform, would go far to solve the grave and serious problem of the fertility of the unfit.

The unfit are subject to no moral law in the matter of procreation. They can be taught nothing, and they will practise nothing. Like the lower animals they obey their instincts and gratify their desires as they arise.

It has been seriously suggested that the poor should be systematically taught Neo-Malthusian methods for the limitation of their offspring.

The best among the poor might practise them, the worst certainly would not, and the limitation among the best would only stimulate the fertility of the worst. This is the most innocent and harmless of the numerous suggestions made by reformers for controlling the fecundity of the poor.

Of surgical methods, castration of males, Oophorectomy or the removal of the ovaries in women, and vasectomy, or the section of the cords of the testicles, have all been suggested.

Annual castration of a certain number of the children of the popular classes was not long ago seriously proposed by Weinhold.

Boies, in his "Prisoners and Paupers," declares that surgical interference is the only method of dealing with the criminal, and preventing him from reproducing his kind. He says:—"These organs have no function in the human organism except the creation and gratification of desire and the reproduction of the species. Their loss has no effect upon the health, longevity, or abilities of the individual of adult years. The removal of them therefore by destroying desire would actually diminish the wants of nature and increase the enjoyments of life for paupers. A want removed is equivalent to a want supplied. In other words, such removal would be a positive benefit to the abnormal rather than a deprivation, rather a kindness than an injury. This operation bestowed upon the abnormal inmates of our prisons, reformatories, jails, asylums, and public institutions, would entirely eradicate those unspeakable evil practices which are so terribly prevalent, debasing, destructive, and uncontrolled in them. It would confer upon the inmates health and strength, for weakness and impotence, satisfaction and comfort for discontent and insatiable desire."

Anæsthetics have ensured that these operations may be performed without the slightest suspicion of pain, and with careful sympathetic surgery, pain may be absent throughout the whole of convalescence. Antiseptics have made it possible to perform these operations with practically no risk to life.

Though castration and Oophorectomy can be performed with safety and without pain, they are absolutely unjustifiable operations, if done to produce sterility.

Every incision and every stitch in surgery, beyond the necessities of the case, are objectionable, and to remove an organ, when the section of its duct is sufficient is to say the least of it, bad surgery.

Vasectomy is the resection of a portion of the duct of the testicles, followed by ligature of the ends. No doubt ligature alone would be sufficient for the purpose, but up to the present, a piece of the duct has been removed, when this operation has been found necessary in the treatment of disease.

This duct is the secretory tube of the testicle, so that when it is occluded, the secretion is dammed back, and degeneration and atrophy of the organ are induced. It soon wastes, and becomes as functionless as though it were removed.

This operation can be performed in a Surgery with the aid of a little Cocaine, and the patient may walk to his home, sterilized for the rest of his natural life, after the complete loss of any accumulated fluid.

Of these two operations for the sterilization of men, vasectomy is preferable. The major operation for the purpose of inducing artificial sterility should never for a moment be considered.

But vasectomy, though surgically simple, and a less violation of sentiment than castration, cannot be justified except in exceptional cases.

Neither of these operations makes the subjects of them altogether or at once impotent, certainly not for years. It sterilizes and partly unsexes them and in the end completely so.

But the physical and mental changes that follow the operation in the young adolescent are grave and serious, and a violent outrage upon the man's nature and sentiment.

Society can hope for nothing but evil from the man she forcibly unsexes; but if he must be kept in durance vile for the whole of his life there is little need for such an operation.

The criminal cases bad enough to justify this grave and extreme measure should be incarcerated for life.

The cases, it has been thought, that fully justify this operation are those guilty of repeated criminal assaults.

Such a claim arises out of insufficient knowledge of the physiology of sex, and the pathology of crime. Emasculation would have little influence in preventing a recurrence of this crime, for the operation does not render its subjects immediately impotent, nor does it change their sexual nature any more than it beautifies their character.

The instinct remains, and the power to gratify it remains at least for some years. With the less knowledge of surgery of earlier times, a social condition in which such a practice might be rationally considered, is conceivable, but with the present state of our profession, such measures would be unthinkable.

CHAPTER XI.

TUBO-LIGATURE.

There is a growing feeling that society must be protected, not so much against the criminal as against the fertility of the criminal, and no rational, practicable, acceptable method has as yet been devised.

The operations on men to induce sterility have been discussed and dismissed as unsatisfactory.

But analogous operations may be performed on women. And if women can be sterilized by surgical interference, whence comes the necessity of sterilizing both?

Oophorectomy, or removal of the ovaries is analogous to castration. It is an equally safe, though a slightly more severe and complicated operation.

It can be safely and painlessly performed, the mortality in uncomplicated cases being practically nil.

The changes physical and mental are not so grave as in the analogous operation on the opposite sex, and they vary considerably at different ages and in different cases. The later in life the operation is performed the less the effect produced. At or after the menopause (about the 45th year) little or no change is noticeable.

In many, and especially in younger women however, grave mental and physical changes are induced. The menstrual function is destroyed, the appearance often becomes masculine, the face becomes coarse and heavy, and hair may appear on the lips and chin. Lethargy and increase of weight are often noticed, and not a few, especially in congenitally neurotic cases, have an attack of insanity precipitated.

On the same principle on which the radical operation on men was condemned, Oophorectomy must also be condemned. It is a serious operation, often attended with grave mental and physical disturbances, not the least of which is the partial unsexing of those subjected to it.

While these are delicate they are also pressing questions, questions which, like the mythical riddle of the Sphynx, not to answer means to be destroyed, yet the sentimental difficulties, are accentuated by modern progress, for the public conscience becomes more sensitive as problems become more grave. But as science has prepared the bridge over which society may safely march, so, with rules easily provided by an enlightened community all remedial measures formerly proposed—wise in their times, probably, may now be waived aside.

With our present knowlege, the simple process of tubo-ligature renders unsexing absolutely unnecessary in order to effect complete and permanent sterility. As the lesser operation vasectomy, is effectual in men, so is a lesser operation, tubo-ligature effectual in women. And it has this paramount advantage that, whereas vasectomy being an occlusion of a secretory duct, leads to complete atrophy and destruction of the testis, ligature of the Fallopian tube, which is only a uterine appendage and not a secretory duct of the ovary, has absolutely no effect whatever on that organ.

A simple ligature of each Fallopian tube would effectually and permanently sterilise, without in any way whatever altering or changing the organs concerned, or the emotions, habits, disposition, or life of the person operated on.

The Fallopian tubes are two in number, attached to the upper angles of the uterus, and communicating therewith. Each is about five inches in length, and trumpet-shaped at its extremity, which floats free in the pelvic cavity.

Attached to the margin of this trumpet-shaped extremity, is a number of tentacle-like fringes, the function of which is to embrace the portion of the ovary, where an ovum has matured during or immediately after menstruation.

At all other times these tubes are practically unattached to the ovaries. Ova may and do mature on the surface of the ovaries, but do not always pass into the Fallopian tubes; being almost microscopic, they are disintegrated and reabsorbed. If they do pass into a tube they are lost or fertilized as the case may be.

It can be seen that the function and vitality of the ovaries are in no way affected by the tubes. The ovarian function goes on, whether the tubes perform their function of conveyance or not, and if this function can be destroyed, life-long sterility is assured. There is no abdominal operation more simple, rapid and safe, than simple ligature of the Fallopian tubes. It may be performed by way of the natural passage, or by the abdominal route, the choice depending on various circumstances. If the former route be taken, there may be nothing to indicate, in some cases not even to a medical man, that such an operation has been performed.

The Fallopian tubes have been ligatured by Kossman, Ruhl and Neuman for the sterilization of women with pelvic deformities; but all testify to the danger of subsequent abnormal or ectopic pregnancy, and several instances are given. Mr. Bland Sutton relates a case in an article on Conservative Hysterectomy in the British Medical Journal.

After numerous experiments on healthy tubes, I have found that simple ligature with even a moderate amount of force in tying will cut the tube through in almost any part of its length. The mucous lining is so thrown into folds that its thickness in relation to the peritoneal layer is considerable. Because of this, the tube when tied alone is brittle, and a ligature applied to it will very easily cut through, and either allow of reunion of the severed ends or leave a patent stump. In a recorded case in which pregnancy occurred after each tube was ligatured in two places, and then divided with a knife, a patent stump was no doubt left.

In order to obviate this danger the peritoneal layer must be opened, and the mucous membrane, which is quite brittle and easily removed, must be torn away for about one quarter of an inch. A simple cat-gut or silk ligature lightly tied would then be sufficient to insure complete and permanent occlusion.

Nature often performs this operation herself, with the inevitable and irrevocable result, lifelong sterility, with no tittle of positive evidence during life of its occurrence.

Here are a few examples:—A young married woman has a miscarriage; it is not severe, and she is indiscreet enough to be about at her duties in a day or two, but within a few days or so she finds she must return to bed, with feverishness and pelvic pain. Before a month is past she is up and quite herself again. But she never afterwards conceives. What has happened? To the most careful and critical examination nothing abnormal is detected. Her general health, her vitality, her emotional and sexual life, her youthful vigorous appearance, all are unimpaired. But she is barren, and why? A little inflammation occurred in the uterus and spread along the tubes. The sides of the tubes cohered, permanently united by adhesive inflammation, and complete and permanent occlusion resulted.

The operation of tubo-ligature is an artificial imitation of this inflamatory process.

Pelvic inflammation, sometimes very slight, following a birth, or the same process set up by uterine pessaries used for displacements, may induce adhesive inflammation in the tubes, and simple and permanent sterility is the incurable result. It is a well known fact that prostitutes are usually sterile, and this arises from the prevalence of venereal disease, which produces gonorrhœal inflammation of the Fallopian tubes, resulting in complete and permanent occlusion.

This process could be best imitated, if cauterisation of the tubes were a safe and reliable procedure. An electric cautery passed along the tubes would result in a simple and speedy occlusion. But in the present state of our gynecological knowledge this appears impracticable.

We have therefore at our hand, a simple, safe, and certain method of stopping procreation by the sterilization of women by tubo-ligature.

This operation would entail no hardship on women. It is so easy, safe and painless, that thousands would readily submit to it to-morrow, to be relieved from the anxiety which a possible increase in their already too numerous families excites. Hundreds of women and men to-day are living unnatural lives, because of their refusal to bring children into the world with the hereditary taint they know courses in their own veins.

Many men are living loose and irregular lives, amongst the easy women of society, because the indiscretion of their youth has damned them for ever with a syphilitic taint, which they could not fail to transmit to their progeny.

Many virtuous men and women are living a life of abstinence from even each other's society, because their physician has taught them something of the law of heredity. Would not all these women readily submit to sterilization?

As it produces no mental nor moral, nor physical change, it violates no law, and outrages no sentiment. It is an outrage upon society, and a greater upon an innocent helpless victim to bring a defective into the world; it is a moral act to prevent it by this means.

And of all the methods yet suggested or devised, or practised, tubo-ligature is the simplest, most effective, and least opposed to sentiment and prejudice.

It will of course be asked:—What about criminals and defective men? Let their wives be sterilized. The wife of any criminal would deem it a boon to be protected from the offspring of such a man, so would society.

If he is not married, then society must take the risk, and it is not very great. The women who will be his companions will be either sterilized by disease or by tubo-ligature, because they are defectives. This protection from the progeny of defective men, though not absolute, is complete enough for all practical purposes.

If all defective women and the wives of all defective men are sterilized, a greater improvement will take place in the race in the next 50 years, than has been accomplished by all the sanitation of the Victorian era.

CHAPTER XII.

SUGGESTIONS AS TO APPLICATION.

The fertility of the unfit goes on unrestrained by any other check, save vice and misery. The great moral checks have not, and cannot have any place with them. But the State is, by its humanitarian zeal, limiting the scope and diminishing the force of these natural checks amongst all classes of the community, but especially amongst the unfit, so that its policy now fosters the fertility of this class, while it fails to arrest the declining nativity of our best citizens. The greater the fertility of the unfit, the greater the burden the fit have to bear, and the less their fertility.

The State's present policy therefore, fosters the fertility of the unfit, and discourages the fertility of the fit. This disastrous policy must be changed without delay. The State can arrest the gradual degradation of its people, by sterilizing all defective women and the wives of defective men falling into the hands of the law. Mr. Henry M. Boies in "Prisoners and Paupers" suggests life-long isolation. He says:—"It is time however that society should interpose in this propagation of criminals. It is irrational and absurd to occupy our attention and exhaust our liberality with the care of his constantly growing class, without any attempt to restrict its reproduction. This is possible too, without violating any humanitarian instinct, by imprisonment for life; and this seems to be the most practicable solution of the problem in America. As soon as an individual can be identified as an hereditary or chronic criminal, society shall confine him or her in a penitentiary at self-supporting labour for life.

"Every State should have an institution, adapted to the safe and secure separation of such from society, where they can be employed at productive labour, without expense to the public, during their natural life. When this is ended with them, the class will become extinct, and not before. Then each generation would only have to take care of its own moral cripples and defectives, without the burden of the constantly increasing inheritance of the past. When upon a third conviction the judicial authorities determine the prisoner to belong to the criminal class, the law should imperatively require the sentence to be the penitentiary for life, whatever the particular crime committed."

M. Boies defines a criminal as one in whom two successive punishments, according to law, have failed to prevent a third offence.

If such a criminal is a woman, she should be offered the alternative of surgical sterility or incarceration during the child bearing period of her life; if a man, his wife should be offered this remedy against the procreation of criminals in exchange for her husband, on the expiry of his sentence, or the protection of divorce.

No woman in the child-bearing period of life should be released from an Asylum, until this operation has been performed. If a man is committed, his wife should have the option of divorce or be sterilized before his release.

A central Board should issue marriage certificates, after consideration of confidential medical reports upon the health, physical condition, and family history of the parties to a proposed marriage contract.

Medical officers should be appointed in the various centres of population by the central Board, and fees on reports should be paid after the manner of Life Insurance fees.

In fact the Life Insurance system would serve as a good model, for the establishment of a system of marriage control, and if questions involving a more detailed family history were added to a typical Life Insurance report form, it could hardly be improved upon, for the purpose of marriage health reports.

If upon consideration of the medical report of the contracting parties, in accordance with the law upon the subject, a certificate of marriage were refused, a certificate of sterilization by tubo-ligature, forwarded to the Board by a Surgeon, should entitle to the marriage certificate.

No law should attempt to step in between two lovers, who have become attached to each other by the bonds of a strong affection, lest a greater evil befall both themselves and society.

A marriage certificate of health should state the complete family history as well as the physical condition of the parties to a proposed marriage, and such certificates should be issued only by the Central Board of Experts, who would receive the medical reports of its own medical officers.

When the principle of artificial sterilization is accepted by the State, the organization necessary to ensure that only the fit shall procreate, will only be a matter of arrangement by experts.

One danger looms ahead however if the operative means of producing artificial sterility are popularised.

Every surgeon of experience knows how readily large numbers of married women encourage surgical treatment for ovarian and even uterine complaints, if they become aware that such treatment is followed by sterility. It is not at all an uncommon thing for women in all ranks of life, to encourage, and even seek removal of the ovaries in order to escape an increase in the family.

They become acquainted with persons who have submitted to this operation for ovarian disease, and noting nothing but improvement in their health, attended by sterility, their intense anxiety to enjoy immunity from child-bearing makes them eager to submit to operation.

It would be distinctly immoral to sterilize healthy women, who become possessed with the old Roman passion for a childless life, or who simply wish to limit their families for any selfish or personal reason.

Any law which recognizes the induction of artificial sterility should make operative interference with those fit to procreate a healthy stock an offence.

Induced sterility should rank with induced abortion, and be a criminal offence, except in certain cases which could be defined.

There is much evidence to suggest that artificial sterilization may become as a great vice, as great a danger to the State as criminal abortion.

Artificial abortion, as commonly performed, is a much more dangerous operation than tubo-ligature. Of the two operations, any experienced surgeon would readily declare that the latter is the simpler and the safer; the one less likely to lead to unfavourable complications, and the one, moreover, that would leave the subject of it with the better "expectancy of life."

Anæsthetics and anti-septics have made this comparison possible and true.

Any surgeon who performs tubo-ligature should be liable to prosecution, unless he can justify his action according to the law relating to the artificial sterility of the unfit.

While the law would eventually require to be obligatory, with regard to the absolutely unfit, it would require to be permissive in all other cases.

Many voluntarily abstain from marriage, because of a strong hereditary tendency to certain diseases such as cancer and tubercle.

There must of necessity be many on the border-land between the fit and the unfit, and clauses permitting sterilization under some circumstances would be required.

CONCLUSION.

In conclusion let us briefly review the whole position taken up in this imperfect study of a great question.

1. The birth-rate is rapidly and persistently declining.

2. The food-rate is persistently increasing.

3. The declining fertility is not uniform through all classes.

4. The fertility of the best is rapidly declining.

5. The fertility of the worst is undisturbed.

6. The policy of the State is inimical to the fertility of its best, and fosters the fertility of its worst citizens.

7. The infertility of the best stock is due to voluntary curtailment of the family, through sexual self-restraint.

8. No such-factor does or can obtain as a check to the fertility of the unfit.

9. The proportion of the unfit to the fit is in consequence annually increasing.

10. The *future* of society demands that compulsory sterilization of the unfit should be adopted.

11. No method ever tried or suggested offers the advantages of simplicity, safety, effectiveness, and popularity, promised by tubo-ligature.

12. The State must protect itself against the collateral danger of artificial sterilization of its best stock.

The highest interest of Society and of the individual urgently requires that the size of families be controlled.

The moral restraint of Malthus (delayed marriage) and post-nuptial intermittent restraint are the only safe and rational methods, that our civilization can possibly encourage, or physiology endorse.

These methods must of necessity be peculiar to the best class of people. For the worst class of people, induced sterility, or prohibited fertility, is an absolute necessity, if Society and civilization must endure.

Now what are likely to be the results of, first, the moral methods, and, second, the surgical method of our curtailment.

"It does not appear to me," says Dr. Billings (Forum, June, 1893), "that this lessening of the birth-rate is in itself an evil, or that it will be worth while to attempt to increase the birth-rate merely for the sake of maintaining a constant increase in the population, because to neither this nor the next generation will such increase be specially beneficial."

To Aristotle, the great advantage of an abundant population was, that the State was secured against invasion by numerous defenders.

If we can find no stronger justification for a teeming population than this to-day, we will be forced to agree with Dr. Billings, that neither to this nor the next generation, is a great increase especially beneficial.

But the moral effect of judicial limitation is very great. If men and women can marry young, one great incentive to vice is removed. If married people can bear their children when they can best support them, they will marry when their bodies are matured, and bear their families when their finances are matured.

For children well provided for, and educated, and born after full physical and mental maturity in their parents, turn out the best men and women.

If the conditions of life are made easy, if ease and comfort are tolerably secured to all, if the strain and stress of life are reduced, if hardship, poverty, and want are reduced to a minimum, the sexual instinct and parental love in human nature, so far unimpaired by any known force, are powerful enough to keep the race alive, and insure a progressive development.

The greater the proportion and the fertility of the defective, the less hope for the future. If the fertility of the unfit be reduced to a minimum, not only will many dreadful hereditary diseases be eradicated, but the fertility of the fit will receive a powerful stimulus, because of the great diminution there will necessarily be in the burdens they will have to bear.

The advantages of sterility to the unfit themselves will, on the whole, be incalculable. They are self-evident, and need not be dwelt on here.

The whole sum of human happiness would in this way be most assuredly increased, and the aim and object of all social reform be to some extent at least, realized.